Alfred Harker

The Bala Volcanic Series of Caernarvonshire and Associated Rocks

Being the Sedgwick Prize Essay for 1888

Alfred Harker

The Bala Volcanic Series of Caernarvonshire and Associated Rocks
Being the Sedgwick Prize Essay for 1888

ISBN/EAN: 9783744749862

Printed in Europe, USA, Canada, Australia, Japan

Cover: Foto ©berggeist007 / pixelio.de

More available books at **www.hansebooks.com**

THE BALA VOLCANIC SERIES

OF

CAERNARVONSHIRE

AND

ASSOCIATED ROCKS;

BEING

THE SEDGWICK PRIZE ESSAY FOR 1888,

BY

ALFRED HARKER, M.A. F.G.S.,

FELLOW OF ST JOHN'S COLLEGE,
AND DEMONSTRATOR IN GEOLOGY (PETROLOGY) IN THE UNIVERSITY OF
CAMBRIDGE.

Sane multum illi egerunt, qui ante nos fuerunt, sed non peregerunt;
multum adhuc restat operis, *multumque restabit, nec ulli nato post mille*
secula praecluderetur occasio adjiciendi.

SENECA.

CAMBRIDGE:
AT THE UNIVERSITY PRESS.
1889

𝕮𝖆𝖒𝖇𝖗𝖎𝖉𝖌𝖊:

PRINTED BY C. J. CLAY, M.A. AND SONS,

AT THE UNIVERSITY PRESS.

PREFACE.

THE following pages are, in substance, the Sedgwick Prize Essay
for the year 1888, the subject proposed by the Examiners
being "The Petrology of the Igneous Rocks associated with the
Cambrian (*Sedgwick*) System of Caernarvonshire." Since the
award, I have completely re-written the essay, and have chosen a
title which more accurately expresses its actual scope.

I am indebted to Professor T. McKenny Hughes for the use of
specimens in the Woodwardian Museum and for kind assistance in
other ways. I have also to thank the other Examiners, Professor
T. G. Bonney and Mr J. J. Harris Teall, for many valuable
suggestions, and Mr E. Hamilton Acton for kindly supplying me
with chemical analyses of a number of Caernarvonshire rocks.

All the specimens described are in the Woodwardian Museum
collections, and to facilitate reference to the original sections
I have given, in square brackets, the numbers of the slides in the
Museum Cabinet.

ALFRED HARKER.

ST JOHN'S COLLEGE, CAMBRIDGE,
October, 1889.

TABLE OF CONTENTS.

MAPS.

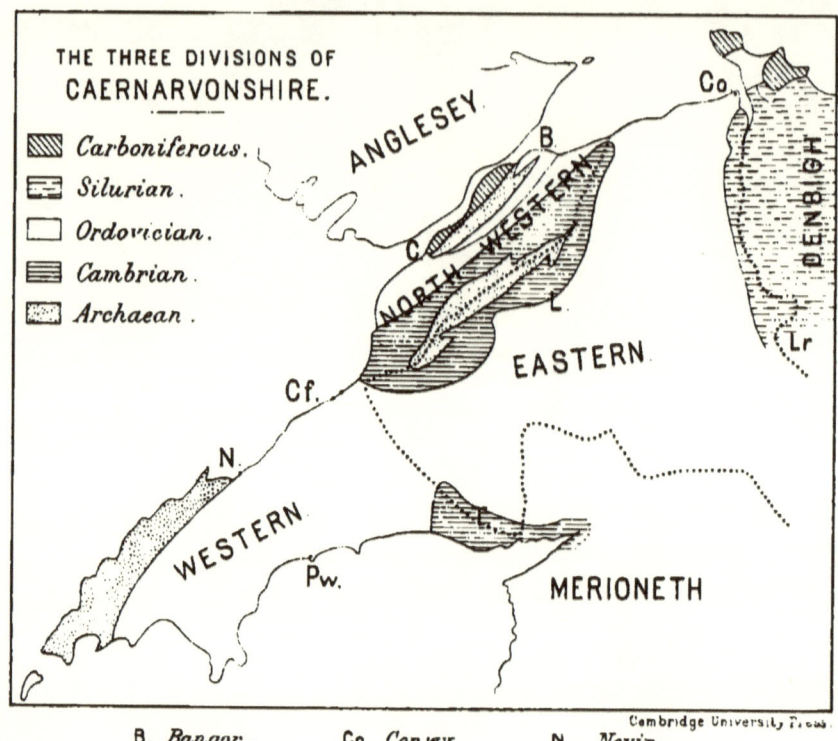

THE THREE DIVISIONS OF
CAERNARVONSHIRE.

- ▨ *Carboniferous.*
- ▤ *Silurian.*
- ☐ *Ordovician.*
- ▤ *Cambrian.*
- ▨ *Archaean.*

Cambridge University Press.

B. *Bangor*.	Co. *Conwy*.	N. *Nevin*.
C. *Caernarvon*.	L. *Llanberis*.	Pw. *Pwllheli*.
Cf. *Clynog-fawr*.	Lr. *Llanrwst*.	T. *Tremadoc*.

Fig. I.

I. Introduction.

The county of Caernarvon, excluding the Great Orme's Head and another small outlying patch near Abergele, has a length of about 52 miles, with a greatest breadth of 20 miles. More than half of its border is washed by the sea, from Portmadoc to Conwy; the valley of the Conwy river forms a natural boundary on the east; while from near the head-waters of that river the county-limit runs along an elevated and sinuous line of water-shed to Llyn Edno, thence westward to Llyn-y-ddinas, and down the river Glaslyn to Traeth Mawr.

The general geology of the county is too well known to require description here, but a few words with respect to its physical features will prepare the way for our subject. Caernarvonshire falls naturally into three divisions of widely different physical character, and in each the scenery is easily correlated with the geological structure and especially with the arrangement of the igneous rocks. (See map, fig. 1.)

The largest division, which we shall name the *eastern*, is limited on the north-west side by the Llyn Padarn ridge, which extends from Llanllyfni to St Anne's Chapel, and which we may imagine prolonged to near Penmaenmawr. Excepting the alluvial flats bordering the Conwy and Traeth-mawr this part of the county presents everywhere a mountainous type of scenery. The mountains culminate in Snowdon and Y Glyder Fawr, and in the Carneddau Dafydd and Llewellyn, the four highest peaks in Wales. The rocks from which the loftiest elevations are carved out consist largely of massive acid lavas, with some tuffs, ribbed by great intrusive sheets of diabase Only in two or three

H. E.   1

localities, as at Penmaenmawr, Bera-mawr and Y Foel Frâs, and
Mynydd Mawr, do large intrusive masses of igneous rocks form
striking features of the landscape.

The *western* division of the county, consisting of the peninsula
of Lleyn, is very different.   Here the hills, though sometimes
rising, as in Yr Eifl, to nearly 1900 feet, owe their imposing
appearance chiefly to relief, and, as compared with the eastern
division, the general surface is level.   The influence on the
scenery of masses of igneous rock, almost all intrusive, is much
more apparent; and in a bird's-eye view of the Lleyn from some
commanding station it is easy to identify each bolder eminence
with its corresponding red patch on the geological map.

The *north-western* part of Caernarvonshire is without either
the great mountains of the eastern division or the abrupt
isolated hills of the western.   With a single exception, it contains no igneous rocks of importance to our subject.

All the igneous rocks alluded to above, whether extruded or
intruded, are to be assigned on good evidence to the Bala period,
and it will be our business in the present essay to describe their
nature and discuss their mutual relations.

To the geologist in Caernarvonshire the admirable maps of
the Geological Survey, supplemented by Sir Andrew Ramsay's
Memoir, are of course invaluable, although they give but meagre
and not always accurate information with regard to the petrology
of the igneous rocks.   The researches of Mr F. Rutley, Professor
Bonney, and Mr Grenville Cole have taught us much with respect
to the rhyolitic lavas of Snowdonia and their secondary modifications; Mr Tawney's *Woodwardian Museum Notes* treat of many
of the intrusive rocks of the Lleyn peninsula; useful contributions
dealing with various other rocks in the county are due to
Mr J. A. Phillips, Professor Bonney, Dr Hicks, and other authors;
and Mr Teall's valuable *British Petrography* affords a large fund
of information on the subject; but the majority of the igneous
rocks of Caernarvonshire are yet undescribed, and such an essay as
the present cannot of course aim at exhausting so wide a subject.

The cabinets of the Woodwardian Museum contain many
rocks collected in Caernarvonshire by Hailstone, Henslow, and
especially Sedgwick, besides those gathered by the late Mr Tawney,
the present Woodwardian Professor, Mr Marr, and others.   The

specimens of the earlier geologists are, however, of little value for microscopic examination, being selected rather to illustrate the appearance of the rock in the field than to reveal its essential characters. They are mostly taken from exposed places such as the tops of hills, and so are usually deeply weathered. Again, it is not always possible to learn the precise localities of specimens obtained by the hand of others, more particularly in the case of the older collectors, who had to work with very imperfect maps, and did not always select rocks *in situ*. For these reasons a personal visit has been made to the home of almost every rock mentioned in this paper, to procure representative specimens from definite spots. This was also necessary in many cases in order to form an opinion on the mode of occurrence of these igneous masses, and their geological relations, often a subject of dispute.

For convenience of reference, a list is given below of the chief books and papers dealing directly with the igneous rocks of Caernarvonshire, so far as they are embraced in the present essay. The abbreviations employed here and throughout the following pages are:

Q. J. G. S. Quarterly Journal of the Geological Society of London.

G. M. Geological Magazine.

Maps of the county on the scale of one inch to a mile are published with geological colouring. Most of the county is included in the quarter-sheets 75 N.E., N.W., S.W., 76 N. and S., 78 N.E., S.E., S.W. I shall assume these maps to be accessible to any one desirous of following the field-geology of the county. The chief horizontal sections are contained in sheets 28 and 31. Since this essay was first written, the whole of the Ordnance Survey of Caernarvonshire on the six-inch scale has been published, and field-geology in that district is now very much facilitated.

Blake, Rev. J. F.

    On the Igneous Rocks of Llyn Padarn, Yr Eifl, and Boduan. *Rep. Brit. Assoc.* for 1886, p. 669 (1887).

    On the Cambrian and Associated Rocks in North-West Caernarvonshire. *Q. J. G. S.*, vol. XLIV., pp. 271—290 (1888).

On the Monian System of Rocks.   *Q. J. G. S.*, vol. XLIV., pp. 463—546 [530—534] (1888).

Bonney, Prof. T. G.

Notes on the Microscopic Structure of some Rocks from Caernarvonshire and Anglesey.   *Q. J. G. S.*, vol. XXXV., pp. 305—308 (1879).

On the Serpentine and Associated Rocks of Anglesey ; with a Note on the so-called Serpentine of Porth-dinlleyn, Caernarvonshire.   *Q. J. G. S.*, vol. XXXVII., pp. 40—50, [48—50] (1881).

On some Nodular Felsites in the Bala Group of North Wales.   *Q. J. G. S.*, vol. XXXVIII., pp. 289—296, pl. x. (1882).

On the so-called Diorite of Little Knott (Cumberland), with further Remarks on the Occurrence of Picrites in Wales.   *Q. J. G. S.*, vol. XLI., pp. 511—521 [517] and pl. xvi. (1885).

On a peculiar variety of Hornblende from Mynydd Mawr, Caernarvonshire.   *Min. Mag.*, vol. VIII., pp. 103—107 and note (1888).

[Notes on Eruptive Rocks in North Wales, pp. 49, 50 in Dr Hicks' *La Géologie du Nord du Pays de Galles ; vid. sub.*]

Cole, Grenville A. J.

On Hollow Spherulites and their Occurrence in ancient British Lavas.   *Q. J. G. S.*, vol. XLI., pp. 162—168, pl. iv. (1885).

On the Alteration of Coarsely Spherulitic Rocks.   *Q. J. G. S.*, vol. XLII., pp. 183—190, pl. ix. (1886).

Elsden, J. V.

Notes on the Igneous Rocks of the Lleyn Promontory.   *G. M.* (3), vol. v., pp. 303—308 (1888).

Harker, Alfred.

Woodwardian Museum Notes : on some Anglesey Dykes, I.   *G. M.* (3), vol. IV., pp. 409—416 [411, 412] (1887).

Notes on the Geology of Mynydd Mawr and the Nantlle Valley.   *G. M.* (3), vol. v., pp. 221—226 (1888).

Additional Note on the Blue Hornblende of Mynydd Mawr.   *G.M.* (3), vol. v., pp. 455, 456 (1888).

On the Eruptive Rocks in the Neighbourhood of Sarn, Caernarvonshire.   *Q. J. G. S.*, vol. XLIV., pp. 442—461 (1888).

Notes on Hornblende as a Rock-forming Mineral. *Min. Mag.*, vol. VIII., pp. 31—34 (1888).

On Local Thickening of Dykes and Beds by Folding. *G. M.* (3), vol. VI., pp. 69, 70 (1889).

Harrison, W. J.

Rambles with a Hammer. The Geology of Criccieth and Pwllheli. *Knowledge*, vol. III., pp. 361—363, 402, 436, 437, 482—484 (1884).

Haughton, Prof. S.

On the Newer Palæozoic Rocks, which border the Menai Straits, in Carnarvonshire. *Proc. Geol. Soc. Dub.*, vol. VI., pp. 1—28 (1854).

Henslow, Prof. J. S.

Geological Description of Anglesea. *Trans. Camb. Phil. Soc.*, vol. I., pp. 359—452 [403, 410—412] (1821).

Hicks, Dr H.

On the Pre-Cambrian (Dimetian, Arvonian, and Pebidian) Rocks in Caernarvonshire and Anglesey. *Q. J. G. S.*, vol. XXXV., pp. 295—304 [298—301] (1879).

La Géologie du Nord du Pays de Galles [pp. 27—50 of *Explications des Excursions* issued in proof to members of International Geological Congress in London, 1888; with sections and coloured map].

Jukes, J. B.

Letters and Extracts from the Addresses and Occasional Writings of; ed. by his sister, London (1871).

Lapworth, Prof. C.

On the Geological Distribution of the Rhabdophora [in eight parts, *Ann. and Mag. Nat. History* (5), vols. III.—VI. (1879—80). Passages bearing on the succession in Caernarvonshire occur vol. III., pp. 245—257; vol. IV., pp. 334, 335, 425, 427; vol. v., pp. 275, 276, 358—360, 366—369].

Marr, J. E.

Fossiliferous Cambrian Shales near Caernarvon. *Q. J. G. S.*, vol. XXXII., pp. 134, 135 (1876).

The Classification of the Cambrian and Silurian Rocks. (*Sedgwick Prize Essay*) Cambridge (1883).

Maw, G.

On a New Section of the Cambrian Rocks in a cutting of the Llanberis and Caernarvon Railway, and the Banded Strata of Llanberis. *G. M.*, vol. v., pp. 121—125, and pl. vii. (1868).

Phillips, J. A.

On the Chemical and Mineralogical Changes which have taken place in certain Eruptive Rocks of North Wales. *Q. J. G. S.*, vol. xxxiii., pp. 423—429 [423—427] (1877).

Raisin, Miss C. A.

On some Nodular Felsites of the Lleyn. *Q. J. G. S.*, vol. xlv., pp. 247—269 (1889).

Ramsay, Sir A.

On the Physical Structure and Succession of some of the Lower Palæozoic Rocks of North Wales and Part of Shropshire. *Q. J. G. S.*, vol. ix., pp. 162—176 (1853).

A Descriptive Catalogue of the Rock-Specimens in the Museum of Practical Geology: London (1858); 2nd ed. (1859); 3rd ed. (1862). [Our references are to the third edition.]

The Geology of North Wales. *Mem. Geol. Surv. Eng. and Wales*, vol. iii.; London (1866); 2nd ed. (1881). [Our references under the name 'Ramsay' are to the second edition of this work.]

Rosenbusch, Prof. H.

Mikroskopische Physiographie der Massigen Gesteine; Stuttgart (1877); 2nd ed. (1887). [Our references are to the second edition.]

Rutley, F.

On Community of Structure in Rocks of Dissimilar Origin. *Q. J. G. S.*, vol. xxxv., pp. 327—340 [335, *etc.*] (1879).

On Perlitic and Spherulitic Structures in the Lavas of the Glyder Fawr, North Wales. *Q. J. G. S.*, vol. xxxv., pp. 508, 509 (1879).

On the Microscopic Structure of Devitrified Rocks from Beddgelert and Snowdon. *Q. J. G. S.*, vol. xxxvii., pp. 403—409, pl. xxi. (1881).

The Felsitic Lavas of England and Wales. *Mem. Geol. Surv. Eng. and Wales ;* London (1885).

Sedgwick, Prof. A.
  On the Classification of the Fossiliferous Slates of North Wales, Cumberland, Westmorland, and Lancashire. *Q. J. G. S.*, vol. III., pp. 133—164 (1847).

Sharpe, D.
  Contributions to the Geology of North Wales. *Q. J. G. S.*, vol. II., pp. 283—316 (1846).

Tawney, E. B. (with Prof. Bonney and A. S. Reid).
  Woodwardian Laboratory Notes; North Wales Rocks. *G. M.* (2), vol. VII., pp. 207—215, 452—458 (1880); vol. IX., pp. 548—553 (1882); vol. X., pp. 17—21, 65—71 (1883).

Teall, J. J. Harris.
  British Petrography; London (1888). [Referred to under the name Teall.]

Trimmer, J.
  On the Alteration produced in a Conglomerate of the Poikilitic Series near the Church of Llanfair Iscaer, Caernarvonshire, by the Contact of a Mass of Trap. *Journ. Dubl. Geol. Soc.*, vol. II., p. 35 (1839).

Waller, T. H.
  Penmaenmawr. *Midland Naturalist*, Jan. 1885, pp. 1—7.

Ward, Rev. J. Clifton.
  Notes on the Comparative Microscopic Rock-structure of some Ancient and Modern Volcanic Rocks. *Q. J. G. S.*, vol. XXXI., pp. 388—420 [401—405, 419] pl. xviii. (1875).

Watts, W. W.
  The Geology of Mynydd Mawr. *G. M.* (3), vol. v., p. 335 (1888).

II. Rhyolitic Lavas.

It is natural to begin by discussing the great group of volcanic rocks, which are so striking a feature in the geology of central and eastern Caernarvonshire, and which may be studied everywhere among the Snowdonian mountains and in the hills between Conwy and Penmachno. There is no need to recapitulate the evidence brought forward by Sedgwick and the many geologists who have followed him to this interesting region, that these acid rocks of the Bala series are true interbedded lava-flows; nor need we do more than acknowledge the minute care with which the mapping of so intricate a country has been carried out by the officers of the Geological Survey.

It does not appear that any attempt has been made as yet to differentiate the rock-types characteristic of different horizons among the lavas of Caernarvonshire, although the maps of the Geological Survey seem to afford a good basis for a systematic study of vulcanicity in this part of Wales. Without proposing to enter upon such an investigation here, it will be well to state summarily the succession and distribution of the several groups of flows. The precise horizon to which each belongs cannot be fixed until the fossil zones have been more minutely traced.

The lowest lavas in the county occur in the south-eastern part, about Migneint and Llyn Conwy. They belong to the Arenig series of volcanic rocks, so largely developed in Merioneth, but thinning out as they enter Caernarvonshire. Certain thin flows of lavas in the neighbourhood of Tremadoc may perhaps represent the westerly termination of the Arenig lavas in this county, but the stratigraphy of this much disturbed tract requires further elucidation. We shall confine ourselves for the most part to the Bala series.

The earliest lavas of Bala age are those of Dwygyfylchi and Y Drosgl. These crop out in a broad, much faulted band from a little south of Dwygyfylchi, a village west of Conwy, to Y Drosgl[1] near Y Foel Frâs, where they are faulted out against the great intrusive mass, and cannot be traced further.

The next group of lavas is that of Pen-yr-Oleu-wen and Carnedd Llewellyn. Four distinct flows are seen in Pen-yr-Oleu-wen, the westerly ridge of Carnedd Dafydd, facing Nant Ffrancon. These merge into two on the southern flanks of Carnedd Llewellyn, and reappear to the south, the upper one dying out, in the anti-clinal dome to the east of Llyn Ogwen.

Above these come the lavas of Y Glyder-fach, Capel Curig, and Conwy Mountain. Three thick bands, uniting to the south, appear in Y Glyder-fach and Y Tryfan. From the latter peak the outcrops run north, curving round and disappearing by attenuation on the south-east flank of Carnedd Dafydd. The same lavas are seen on the east side of the anticlinal dome above mentioned, and may be traced north by Ffynnon Llugwy, past the east side of Y Foel Frâs, reappearing, broken by faults, in Foel-llwyd and other hills to the north. The mass forming Conwy Mountain probably belongs to the same group, though the mapping here is not quite satisfactory. Along this northerly stretch only one flow, or continuous series of flows, can be distinguished. Lavas belonging to this group appear beyond the Cefn-y-capel syncline, forming a ring around the Capel Curig lakes. On the north-west side three flows are divided, but on the opposite side there is only one, and this thins away immediately east of Capel Curig.

We come next to the main Snowdon lavas. The lower parts of Snowdon and Moel Hebog are carved out of a thick, uninterrupted series of lava-flows, and the same rocks are seen everywhere about the Gwynant and Glaslyn valleys from Pen-y-gwryd to beyond Pont-Aberglaslyn, and also form a triangular patch seven miles in length in the plateau of Llwyd-mawr. In Clogwyn-dur-Arddu and the Pass of Llanberis the series is sometimes divided by slates into five or six flows or sets of flows, and the lowest of these is

---

[1] Two miles north of Foel Frâs : there is another hill of the same name two miles west of that mountain.

seen as far as Carnedd Dafydd. Outliers occur in the synclinal trough west of Dolwyddelen, and also near Moel Siabod, others in Glyn Lledr and near Bettws-y-coed; while a complex set of lavas and ashes belonging to this group occupies most of the ground between the last-named place and Llyn Cawlyd, and northward to Llanbedr. With diminishing thickness these rocks reappear near Llangelynin, and the lavas which build the abrupt hills near Diganwy, beyond the Conwy estuary, seem to represent the final thinning out in this direction of this group of volcanic rocks.

The main lavas in Snowdon, Nant Gwynant, Moel Hebog, and the Dolwyddelen syncline are succeeded by the marine calcareous ashes which have yielded fossils at the summit of Snowdon and in other localities. Above these in a few places are left small outlying patches of an upper lava-flow. Its thickness, as seen above Twl-du at the head of Cwm Idwal, does not appear to be great, and neither it nor the subjacent calcareous ashes are recognisable in the eastern part of the county.

Having regard to the chemical and mineralogical constitution of all these lavas, it appears that most of them are of thoroughly acid characters, and may with propriety be spoken of as rhyolites.

None of the groups of lava-flows enumerated above can be traced westward of the Llwyd-mawr plateau, but certain rocks of allied nature are found in parts of the Lleyn district. Unfortunately the relations of these rocks are often obscure. The maps of the Geological Survey mark with one colour, under the names 'felspar-porphyry' and 'intrusive felspathic rocks,' quartz-porphyries and rhyolites, both intrusive and interbedded, besides occasional patches of acid agglomerates and of andesites, and large tracts of granite-porphyry and granophyre. These will be separated, so far as is possible, but much more detailed mapping is necessary to elucidate completely the relations of the igneous rocks in this western portion of Caernarvonshire. Some interesting transitions from the rhyolites to the rocks of the plutonic bosses will be referred to in the description of the latter. For the present it is sufficient to remark that the descriptions to be given of the rhyolites of the east apply generally to many isolated patches of similar rocks met with in the west. There is, however, this distinction, that while the eastern rhyolites are without exception true lava-flows, those of the western area are in many cases small intrusive masses.

Nevertheless truly bedded rhyolitic lavas certainly occur in several parts of the peninsula, and may perhaps be explained to some extent as a series of flows, not continuous but lying about one or two definite horizons. In the tract immediately north of Pwllheli (see map, fig. 3) there appear to be two massive flows, coalescing to the north-east and to the south-west, with tuffs and ashes at two or three horizons, as may be seen near the town itself. These lavas are probably broken by intrusions of similar rocks, as they are also by three laccolites of diabase. The strike of this volcanic series is a little curved with a general dip to the north-west. Again, from near Llanbedrog a broad tract of rhyolitic lavas stretches north-north-westerly, curving a little towards north and north-east and apparently having a general dip to east-north-east, east, and east-south-east successively. Agglomerates and ashes occur in this series at 'Pig Street' (Llanbedrog) and Y Gledrydd (Madryn), having an east-south-easterly dip at the latter locality (see map, fig. 2). A large red patch appears here on the Survey Map, but some part of the country so coloured is certainly occupied by intrusive acid rocks. The extreme corner of the tract at Crugau near Llanbedrog is of intrusive rhyolite, and the large mass to the south of the last-named place is of granophyric quartz-porphyry, also intrusive: the long tongue crossing Nant Bodlas appears to be a sheet of quartz-porphyry intruded along the bedding of the shales. Some of the rhyolites about Carn-guwch to the south of Yr Eifl may be interbedded flows, and perhaps these, with the lavas just before mentioned and those of Pwllheli, may belong to a single series, holding a position not very different from that of the main Snowdon lavas. It should be noted that, so far as can be conjectured in a district so laden with superficial deposits, this central part of Lleyn appears to be an elongated synclinal basin, its axis extending from somewhere between Boduan and Llanbedrog in a north-easterly direction. The syncline dies out to the north-east, and is probably broken on its north-westerly side.

Some three miles east of Pwllheli occurs a large patch of rhyolite, partly bedded, and associated in the Pen-y-chain headland with coarse and fine agglomerate [1]. The lavas are apparently complicated in the northern part by intruded rhyolites, but in the main belong to an interbedded group. The overlying shales at Pont-llym-gwyn dip north-westerly, and this group of lavas thus appears to pass under those of the Pwllheli neighbourhood. They are doubtless prolonged under the sea in a general south-westerly direction, though whether they die out or pass into the Llanbedrog lavas it is impossible to say. The Careg-y-defaid rhyolite, between Pwllheli and Llanbedrog, seems to be intrusive. The same is true of the rhyolite at Broom Hall, east of Pwllheli, forming the southern part of the red tract on the map (the northern portion is diabase). Other rhyolitic rocks which apparently have intrusive relations with the surrounding strata are those at Criccieth Castle, Plas-hên, and Mynydd Edynfedd, and the deeply-weathered mass forming Carn-penteyrch near Llangybi, mapped by the Survey as greenstone. Here I have not included some other rocks which will be spoken of later as rhyolitic quartz-porphyries.

The earliest microscopical notice of the Caernarvonshire lavas seems to be that of the late Mr J. Clifton Ward [2]. This writer endorsed the views already expressed by Sir A. Ramsay as the result of the Survey's investigations, and endeavoured to draw a parallel between the 'Snowdon felstones' and modern volcanic rocks.

The view gained ground in England, though for the most part disregarded by continental petrologists, that the Bala lavas of Caernarvonshire differ in no essential character from Tertiary and Recent acid flows, the differences they do exhibit being the results of secondary changes, and especially of the devitrification of an originally glassy mass. Mr Rutley [3] discovered the remains of undoubted perlitic structure in one of the "felstones" of Esgair-felen, near Y Glyder Fawr, while others from the same neighbourhood afforded examples of the spherulitic and fluidal structures. The same author subsequently described and figured other spherulitic and perlitic rocks from near Beddgelert and

[1] Described by Miss Raisin. Mr Harrison's suggestion of a pre-Cambrian age for this tract is one for which I can find no warrant whatever.

[2] Q. J. G. S., vol. xxxi., pp. 399 et seqq., and pl. xviii.; 1875.

[3] Ibid., vol. xxxv., p. 508; 1879.

from the Pass of Llanberis[1], which he regarded as devitrified obsidians and rhyolites; and still further illustrations are given in his memoir on the *Felsitic Lavas of England and Wales*[2]. The researches of Professor Bonney, Mr Grenville Cole, and Miss Raisin on the curious nodular structures so common in these rocks will be mentioned in their proper place. Mr Teall has devoted some notice to the Welsh lavas in his *British Petrography*.

Unfortunately we have very few chemical analyses of the Caernarvonshire lavas. Four are quoted below, the second one having been kindly made for me by Messrs Acton and Hewitt. All these are from the main Snowdon lavas. Nos. I. and II. may be regarded as fresh examples from the lower part of this group; the other two are from nodular varieties, which give evidence of secondary changes. The last, indeed, is the white matrix of the nodules, and Mr Cole considered some part of the silica as probably secondary.

	I.	II.	III.	IV.
SiO_2	74·88	77·5	79·72	83·08
Al_2O_3	12·00	9·7	9·65	10·25
Fe_2O_3	3·50	6·1	5·69	trace
FeO	0·20	not est.	—	—
CaO	0·34	—	not est.	0·26
MgO	1·28	—	not est.	0·09
K_2O	4·77	5·8	1·93	1·78
Na_2O	2·49	0·3	3·54	3·58
Loss on ignition	1·20	0·4	—	0·74
	100·66	99·8	100·53	99·78

I. Pitt's Head. Anal. Hughes; *Trans. Roy. Ir. Acad.*, vol. XXIII., p. 615.

II. Cwm-silyn, Nantlle. Anal. E. Hamilton Acton and J. T. Hewitt.

III. Lledr valley. Anal. F. H. Hatch; *Q. J. G. S.*, vol. XXXIX., p. 485.

IV. Digoed. Anal. G. A. J. Cole; *Q. J. G. S.*, vol. XLII., p. 187.

[1] *Q. J. G. S.*, vol. XXXVII., p. 403, and pl. xxi.; 1881. .
[2] *Mem. Geol. Surv.*; 1885.

On comparing the four analyses, which stand, so far as I can
make out, in ascending order of succession, we note a progressive increase in the silica-percentage. This is such as cannot be explained
quite satisfactorily by alteration having occurred in the nodule-
bearing rhyolites, although in the case described by Mr Cole
the action has been such as to raise the percentage of silica in
the white part of the rock. The falling off in the total alkalies
in analyses III. and IV. is, however, most probably accounted
for by the secondary action experienced, the material abstracted
from the matrix and leached into the shell-like dark layers of
the nodules having 6·21 *p.c.* of potash and 2·19 of soda. This
may also explain partly the excess of soda over potash in columns
III. and IV.

To obtain some idea of the interpretation of the analyses,
I have selected the Pitt's Head rock (I.) and calculated the proportions of the several constituents seen in the thin sections.
These are quartz, orthoclase, albite, magnetite, and a greenish
substance which in the hand-specimens is black. It is necessary
to make some assumption as to the composition of this last
substance, and I have accordingly assumed it to be the same
as the "pinite" analysed by Mr Cole in the Digoed rhyolite, a

	Quartz.	Orthoclase.	Albite.	"Pinite."	Magnetite.	Difference.	Total = Bulk-analysis.
SiO_2	40·03	14·78	12·61	7·46	...	...	74·88
Al_2O_3	...	4·23	3·61	4·16	...	...	12·00
Fe_2O_3	...	...	...	0·53	0·44	+ 2·53	3·50
FeO	...	...	...	...	0·20	...	0·20
CaO	...	...	...	0·38	...	− 0·04	0·34
MgO	...	...	...	0·27	...	+ 1·01	1·28
K_2O	...	3·86	...	0·91	...	...	4·77
Na_2O	...	...	2·17	0·32	...	...	2·49
Loss	...	...	...	0·64	...	+ 0·56	1·20
	40·03	22·87	18·39	14·67	0·64	4·06	100·66

supposition which is found to give a good result. The figures in the difference column may be taken to indicate that the "pinite" substance of Pitt's Head contains a little more magnesia, iron, and water than that of Digoed, and the mineralogical analysis of the rock is seen to be very nearly:

Quartz	41
Orthoclase	24
Albite	19
"Pinite"	15
Magnetite	1
	100

It is to be regretted that we have not for comparison any analyses of the much-neglected rhyolites of Westmorland, which bear the closest resemblance in the field and under the microscope to these Snowdonian lavas, and belong to the same age. The Bala lavas of Wicklow and Waterford differ in structure from those of Caernarvonshire, and Dr Hatch[1] has shewn that they are extraordinarily rich in soda and comparable with the 'keratophyres' of the German petrologists. His analysis is given below (V.).

It is not difficult to match the Snowdon lavas among those 'younger' rocks to which the Continental geologists restrict the name rhyolite. I select a rock from Custer County, Colorado, the description of which by Mr Whitman Cross[2] shews many points in common with the Caernarvonshire rhyolites (VI.). Its similarity in chemical composition to the rocks I. and II. will be evident, and we shall have occasion more than once to refer to similarities of structure also.

The Arenig lavas of North Wales will not be discussed here. They have a considerable range of chemical constitution, as will be seen on comparing the analyses VII. and VIII. The former, from Aran Mowddwy, probably has secondary quartz, as appears to be the case in many of the higher lavas of the Arenig series [698]. It compares rather closely with the altered Digoed rock (IV.).

[1] G. M., 1889, pp. 70—73.
[2] Proc. Colorado Sci. Soc., 1887.

	V.	VI.	VII.	VIII.
SiO_2	77·29	75·20	83·802	68·8
Al_2O_3	}14·62	12·96	7·686	14·9
Fe_2O_3 trace		0·37	0·111	0·9
FeO	—	0·27	0·408	4·3
MnO	—	0·03	—	—
CaO	trace	0·29	0·896	1·9
MgO	0·38	0·12	0·109	1·1
K_2O	0·16	8·38	2·161	2·8
Na_2O	7·60	2·02	4·229	2·7
P_2O_5	—	trace	0·089	—
Loss on } ignition	0·57	0·58	0·301	2·0
	100·62	100·22	100·000	99·4

V. Brittas Bridge, Wicklow. Anal. F. H. Hatch; *G. M.*, 1889, p. 72.

VI. Custer County, Colorado. Anal. L. G. Eakins; *Proc. Colorado Sci. Soc.*, 1887.

VII. Aran Mowddwy. Anal. J. Hughes; *Q. J. G. S.*, vol. XXXI., p. 400: (with FeS_2 0·191, SO_3 0·017, and CO_2 trace).

VIII. Arenig. Anal. J. H. Player; Teall's *Brit. Petr.*, p. 339.

We shall describe first the normal type of the rocks, and note afterwards the more common modifications. Though exhibiting some considerable variety of aspect in the field, owing to the usual weathering processes, the common type of the Snowdonian lavas is, in fresh specimens, extremely uniform in general appearance and in essential characters. It is a black or dark iron-grey rock of compact aspect and sub-conchoidal fracture, with scattered crystals of glassy felspar shewing, in the freshest examples, the characteristic striation of plagioclase under a hand-lens. By weathering, the porphyritic felspars lose their lustre and become dull white, while the colour of the ground-mass passes through lighter shades of grey to white and yellow, or sometimes pink. Partial decomposition renders the rock less compact, and often brings out a laminated structure following in some cases the direction of flow, in others an incipient cleavage. The porphyritic crystals are commonly quite small and not very closely strewn through the matrix.

The fluxion-structure is often beautifully exhibited on weathered surfaces in the field : it is well seen at many points in Nant Ffrancon and the Pass of Llanberis, on the hills near Conwy, and on Llwyd-mawr.

Distinguished from the occasional imperfect cleavage-structure there is often a fine platy jointing, which has a strong tendency to run with the direction of fluxion, though not following any marked sinuosities in the flow. Columnar structure is seen near lakes Llydau and Idwal, at Cwm-silyn above the Nantlle valley, and in other places.

The dark colour of the rocks in fresh specimens is probably due to the chloritoid substance which the microscope shews to be disseminated through the mass in specks, shreds, and streaks. There is a curious variety of lava in which black flakes and patches, often an inch or two in length, lie in a grey matrix. The patches are flat and parallel to the surface of flow. In thin sections they are seen to consist of aggregations of apparently the same chloritoid mineral. That they are not flattened vesicles filled by secondary infiltration is sufficiently proved by their irregular shape and indefinite boundaries, and by comparison of the slides with those cut from undoubtedly vesicular varieties. Good examples occur in the Diganwy hills and in the road at Llam Trwsgyl near Cae-gors, about two miles north of Beddgelert. Instead of definite patches, there are sometimes thin films of a greasy-looking greenish-black colour.

Some of the rocks exhibit to the eye little streaks and lenticular veins of quartz, still following the lines of flow. These will be discussed below. In a few places this peculiarity is very striking, and, combined with the dark films just mentioned, gives the lava a general resemblance to a gneiss or coarse mica-schist[1]. The finest instance is seen at the base of the main Snowdon lavas where they are crossed by the Beddgelert and Caernarvon road : sections are exposed on both sides of the synclinal trough, near Glan-y-gors and between Pitt's Head and Llyn-y-gader. Here the lenticular streaks of quartz vary up to a quarter of an inch in width. The larger ones enclose the little porphyritic felspars, while the smaller wind round them in the usual fluxional manner. These points and the irregular boundary of the streaks, which are

[1] Cf. Bonney, Q. J. G. S., vol. xxxviii., pp. 289, 290 ; 1882.

often drawn out in comb-like processes at their terminations, prove that here also we have not to deal with filled-in cavities in the rock.

Omitting for the present the spheroidal and nodular varieties, we proceed to the microscopic examination.

Many of the slides[1] shew no minerals except felspars and quartz, with usually some magnetite in dust or little granules. Sometimes this last mineral is rather abundant [604], and in some cases it forms good crystals [656, 678, 854], while the fine dust may be seen following the lines of flow in the slice. It is quite possible that some of this dust may be a secondary product, and in one or two cases there is a quantity of secondary magnetite collected within the porphyritic felspars, the source of which is not very evident [33]. The true Snowdon rhyolites are those poorest in magnetite.

Some ferro-magnesian mineral seems to have played a subordinate part in almost all the rocks studied, but in most cases it has been totally destroyed, and is represented only by pale 'viridite.' The main Snowdon lavas are perhaps poorer in evidence of such constituents than the lower groups of flows. A few grains of colourless augite are seen in some of the Y Glyder Fach specimens and in a rock from Llyn Ogwen [311], as well as in one or two of the Lleyn examples. It seems probable that the pale green product so common in the slides results from the destruction of this mineral.

A specimen from near the top of Y Glyder Fach [655] has clusters of green biotite flakes, the colours, which probably indicate some degree of alteration, being for vibrations

parallel to the cleavage traces   —intense brownish green,
perpendicular to cleavage traces—rather pale yellow.

Other slides shew flakes of a green mineral with pretty strong dichroism and moderate double-refraction, representing no doubt a more altered mica. Rarely [245] there is a brown dichroic substance with no definite structure, which may possibly be another form of altered biotite. No clear evidence of hornblende is found in any of the rocks.

---

[1] About sixty specimens from various parts of the county have been sliced and examined.

The commonest decomposition-product, already mentioned, is pale green to colourless and very feebly polarising or sensibly isotropic. It forms in some cases patches suggestive of pseudomorphs, but generally occurs in specks, streaks and lines following the direction of flow, and, when the fluxion is tortuous, collecting in the loops of the curves. This phenomenon and the similar arrangement of the streams of magnetite dust seem to indicate a certain permeability of the rocks in the direction of flow. The properties of the 'viridite' material connect it rather with the chloritoid than the chlorite family[1]. It is perhaps the same substance as the dark patches noticed above in the hand-specimens, and may be delessite; but here chemical analyses are a desideratum, for the black substance in some of the spheroidal rhyolites has been proved by Mr Grenville Cole to have a composition near that of the so-called 'pinite,' and we have already seen in the case of the Pitt's Head rock that the coloured constituent there is one rich in alkalies. I can find no account of the optical properties of 'pinite.'

Some slides have a colourless or yellowish doubly-refracting mineral which forms thin films filling cracks in the direction of flow. It seems to be connected with some crushing of the rock-mass, and is perhaps the sericite of Professor Bonney's paper.

In one section, from the top of Llanberis Pass [650], are rows of brownish, highly-refractive, brightly polarising grains, which must be referred to sphene.

The lavas of the eastern division of Caernarvonshire are not rich in porphyritic elements. This is especially the case in the main Snowdon group, in which, as a rule, only a few scattered felspars occur, and quartz grains are rare [648]. The Glyder Fach group of lavas has rather more abundant felspars [656], but still little porphyritic quartz. The Lleyn rhyolites are more uniformly porphyritic, having often plenty of idiomorphic quartz, corroded and with inclusions of the ground-mass [32, 606, 680, 707], besides felspars which sometimes contain cavernous inclusions of the surrounding matrix [680]. Although these are found in the lava-flows of the Pwllheli district as well as in the little intrusive bosses, they are among the characters which indicate a departure

---

[1] Cf. Heddle, "Chapters on the Mineralogy of Scotland," *Trans. Roy. Soc. Edin.*, vol. xxix., p. 55; 1879.

in this south-western division of the county from the normal rhyolites of the east, and an approach to the rocks described below as quartz-porphyries.

The quartz-grains, when present [648], contain both fluid- and glass-cavities.

The felspars shew a rectangular, squarish, or rather irregular outline, and are usually about a tenth of an inch in length. They almost invariably have the twin-lamellation of plagioclase, often combined with Carlsbad [655, 675] and sometimes with pericline-twinning. In a specimen from near Glyder [657] the crystals are crowded with minute colourless needles of a mineral which I am not able to identify. Similar microlitic needles occur also in the ground-mass of the rock, as at Craig-cwm-silyn near Nantlle [808]. The felspars, by their extinction-angles, may be referred to the albite-oligoclase series. The small percentage of lime found in the analyses shews indeed that they must be of a distinctly acid variety.

Occasionally (near Glyder [657]) we find a crystal broken and drawn slightly apart in the direction of flow, the spaces being filled in with clear quartz-mosaic, a point of some significance.

The ground-mass of these rhyolites shews in almost every case clear indications of fluxion-structure, which follows lines sweeping round the imbedded crystals, and sometimes curving in a very tortuous fashion. Bands of material of slightly different natures often emphasise the lines of flow, and the structures of the ground, both original and secondary, have a marked tendency to follow the same direction.

The perlitic structure is sometimes well marked, as has been recorded by Mr Rutley. It runs frequently in particular bands, and these may have a confusedly crystalline structure, which, taking the perlitic cracks as evidence of an originally glassy mass, indicates secondary devitrification [605, 621, etc.].

An imperfect micro-spherulitic character affects some of the rocks studied, but only occasionally in such perfect development as to give a definite black cross under the polariscope. Specimens from Pitt's Head, Capel Curig Hotel, and the Lleyn district may be cited [653, 675, 604, 621]. The last-named shews in addition a very delicate micro-pegmatite growth, but this seems to be a rare occurrence. More frequent instances of spherulitic structure may be found, however, in the altered rocks, and it has probably been obliterated in many instances by secondary processes. We shall find the same fact to hold good in the coarsely spheroidal or nodular rocks, which do not differ essentially from these micro-spherulitic varieties.

The most usual nature of the ground-mass in the normal rhyolites is the crypto-crystalline ('micro-felsitic'), passing sometimes into a confusedly micro-crystalline structure. This is sometimes uniform throughout the slide, but in other cases it is associated with streaks, in the direction of flow, which have a structure rudely approximating to the axiolitic, and consisting of wedge-shaped or fan-shaped growths of quartz and felspar, the former predominating, set perpendicular to the lines of flow. Where the fluxion is sinuous, the loops of the sharper bends sometimes present a more spherulitic appearance than the rest of the streaks, but the orientation of the elements is never very pronounced [648, etc.].

Allied to this structure are the patches, streaks, and lenticular veins of micro-crystalline material which are frequent in the crypto-crystalline ground of some varieties. They are drawn out in the direction of flow, and have no very definite boundary against the surrounding mass. They too consist mainly, and often entirely, of quartz, which forms a mosaic of clear grains including both fluid- and glass-cavities. One is disposed at the first glance to consider these lenticular patches as indicating a secondary recrystallisation, but the results of a more careful study decidedly negative this idea. The structure seems to be due to an original segregatory process, marking the latest phase in the consolidation of the liquid lava. It has perhaps a certain resemblance to the so-called eutaxitic structure, but finds a closer parallel on a larger scale in the acid ' segregation' or 'contemporaneous' veins of certain crystalline rocks.

The most striking example of this structure is seen in remarkable quartz-streaked rock already noticed as occurring near Pitt's Head [1] on the Caernarvon and Beddgelert road [652, 653].

---

[1] The exposure is about 150 yards north of the rock named Pitt's Head. The Pitt's Head rhyolite, of which an analysis has been quoted above, occurs to the south of that locality, and is an ordinary crypto-crystalline variety.

Here the lenticular patches, close-set in the crypto-crystalline matrix, consist of a spherulitic border passing, sometimes through a very minute micro-pegmatite, into a micro-crystalline mosaic in the interior of the patch. The spherulitic portion has an arrangement partly linear, and partly centric, the aggregate polarisation being more or less distinct at different points, and sometimes exhibiting a very perfect black cross between crossed Nicols. The crystalline aggregate which constitutes the inner portion of the lenticles consists of grains of clear quartz, with minute brownish-yellow glass-cavities, and felspar in very subordinate quantity[1]. The glass-cavities, the admixture of felspar grains, and the gradual passage into the crypto-crystalline matrix, with the spherulitic character of the marginal part of the lenticles, prove that we are not dealing here with the results of any secondary process of the nature of infiltration or recrystallisation, and we must therefore recognise these highly quartzose lenticles as marking a new type of original structure, which, in various degrees of development, is not infrequently met with in the Snowdonian lavas.

It must be confessed, however, that in many instances among the slides examined from the Caernarvonshire rhyolites it is difficult to discriminate between primary and secondary quartz. Where the latter occurs in veinlets traversing the rock [675, *etc.*], or when it evidently replaces felspar crystals [311, 854, *etc.*], it is of course unmistakable; but in other cases the action which gave rise to the separation of secondary quartz seems to have taken effect along the convoluted lines of flow, and especially in their loops, and the appearance thus produced is less easy to distinguish from an original structure [650, 657].

Another well-marked type of structure deserves special notice. It is frequent in the lowest Bala lavas at Y Drosgl near Dwygyfylchi [854] and especially about Penmaenbach[2] [324, 742], but it is also seen in the Y Glyder Fach set of lavas on Foel-llwyd [695], in the main Snowdon lavas at Craig-cwm-silyn [808], and

---

[1] Bands rich in quartz are noticed by Mr Whitman Cross in the rhyolite from Custer Co., Colorado, which, as shewn above, is closely similar in chemical composition to the rhyolite of Pitt's Head. *Proc. Colorado Sci. Soc.*, 1887, p. 230.

[2] The Penmaenbach rock is here referred to the lowest set of the Bala lavas, although the faults which bound it leave the point uncertain. It is mapped by the Survey as intrusive, but the evidence for this view does not appear.

in some of the rhyolites of the Pwllheli district [680]. In ordinary light the ground-mass has a micro-felsitic appearance, but is closely studded with round or elliptic light spots, about 0·02 inch in diameter, each marked out by a dusty-looking border. On using polarised light, it is seen that within these little areas the quartz and felspar are more or less completely individualised, the former being in excess of the latter, and that the whole of the quartz in one of these spots behaves as one crystal. When the structure is imperfectly developed, the spots are of rather irregular form [695], but in the best specimens [808] they are sharply defined and elliptic in outline, while the opaque-looking dust, as if eliminated from the interior, forms a narrow elliptic ring just within the margin of the spot.

The structure just described seems from appearances to be an original one. It has no connection with spherulitic growth, but seems to be rather of the nature of an ophitic structure, the quartz enclosing in part the felspathic constituent. This is well seen when, as is sometimes the case, the felspar occurs in the form of minute microlites with a partial fluxional arrangement [324].

The tendency to a microlitic separation of the felspar is seen in a few of these Penmaenbach rocks [812], and in one or two other localities, as at Four Crosses near Pwllheli [664].

The various types of ground-mass enumerated above include most of those seen in the acid lavas of Caernarvonshire. The highest lavas of the district, which remain only in a few outliers about Snowdon, Nant-Gwynant, and Moel Hebog, I have not had the opportunity of studying. A specimen from Llyn-y-cwn appears in the section [853] to be largely made up of little felspar prisms, and has a more trachytic look than any of the other rocks examined.

A vesicular structure is not specially characteristic of the Caernarvonshire lavas, but it is met with in some of the less massive flows. It occurs, for example, in the Pen-yr-Oleu-wen rhyolites, which are rather low down in the Bala Volcanic Series. A specimen [699] shews that the rock, unlike most of the other lavas, has abundant porphyritic crystals, including quartz and orthoclase as well as the usual albite and magnetite. The ground-mass shews sinuous lines of flow, which wind round the vesicles.

These are very minute and closely set, spheroidal in outline, and either occupied by dirty-looking quartz full of fluid-cavities or lined with quartz and filled in with calcite.

Another good instance is the rock mapped as 'felspathic ash and volcanic conglomerate' about a mile and a half south-east of Llangelynin in the Conwy district. This is a rhyolite belonging to the main Snowdon group. Its vesicles are ellipsoidal in form, 0·02 to 0·1 inch in length, and always drawn out in the direction of the curving lines of flow [696]. Some are lined with quartz and filled in with a pale-green substance of the chloritoid family, polarising feebly in indigo and neutral tints; others have a lining of pale-green, moderately polarising fibres set perpendicular to the wall, while the interior is occupied by confusedly granular quartz; in others again the green border is wanting. In this example the vesicles constitute the bulk of the mass.

The most striking example of a vesicular lava, however, is seen in some rocks, probably faulted outcrops of one small flow, in the Tremadoc district. They are exposed at Ynys-tywyn (not coloured on the Survey map) and at Ynys-cerig-duon (mapped as 'intrusive felspathic trap'); while the little ridge to the north of Ty-obry on the other side of the Traeth (coloured as 'felspathic ashes and volcanic conglomerate') is perhaps a continuation of the same flow. This lava probably underlies the whole of the flows which have been referred above to the Bala Volcanic Series, and further investigation may prove it to be a prolongation of the Arenig lavas. The rock seems to be a trachyte impregnated with secondary quartz and calcite, but is too decomposed for any further specification to be possible [679]. It is remarkable in the field from the occurrence in the fine-grained grey matrix of numerous ellipsoidal amygdules up to two inches or more in length, elongated in the direction of flow. Some are so drawn out as to have the form of narrow cylindrical tubes. They are mostly lined with quartz and filled in with calcite, or entirely occupied by the former mineral. The larger ellipsoidal ones are sometimes lined with good crystals of quartz, upon which the calcite has been subsequently deposited.

It may be noted that a few quartz-filled amygdules are sometimes met with in the small intrusive or seemingly intrusive masses of rhyolite in the Lleyn peninsula, as, *e.g.*, at Careg-y-defaid [700].

To the fragmental and quasi-fragmental rocks of the acid volcanic series only a summary notice will be accorded. True ashes, volcanic agglomerates, and breccias are of much less extensive occurrence than is generally supposed. Large tracts of country so mapped by the Geological Survey, especially about Diganwy, Dwygyfylchi, Llangelynin, Llanbedr, Llyn Crafnant, and Cwm Tryfan, are in reality composed almost exclusively of solid rhyolitic lavas. There are, of course, the well-known fossiliferous tuffs or ashes of Snowdon and Moel Hebog, which consist of volcanic ejectamenta and marine sediment in variable proportions, and small accumulations composed solely of volcanic fragments and dust occur among the true lavas in various parts of the country, e.g. on the Glyders, where they are not mapped; but, as a whole, the Bala Volcanic Series of Caernarvonshire is rather remarkable for the paucity of genuine ashes and agglomerates.

Probably the nodular rhyolites to be described below have been mistaken in some instances for agglomerates, and it is evident that many fine-textured rocks have been set down in the field as 'felspathic ashes', when a microscopical study would have revealed their true character of lavas. Undoubtedly there are cases in which, owing to the supervention of secondary changes, it is very difficult to determine whether a particular hand-specimen or a single rock-slice should be regarded as a lava or as an ash; but in the writer's opinion this difficulty has been very much exaggerated by Mr Clifton Ward[1], and the large majority of the rocks which he refers to metamorphosed ashes are veritable rhyolites. The process of alteration which so frequently masks the true character of these rocks seems to be ordinary decomposition, with the production of secondary quartz, rather than metamorphism in the ordinary sense of the word. One criterion, which I have found useful, lies in the fact that in true fragmental rocks in which crystals and fragments of crystals (usually the former) are imbedded in a fine matrix, though the latter may shew a pronounced lamination, even more marked than the fluxion-structure of the lavas, the included crystals do not share in this general orientation, as they would in a true rhyolite, but lie at all angles to the direction of lamination [656].

[1] Q. J. G. S., vol. xxxi., p. 388; 1875.

Again, a slide which on a general view presents a thoroughly fragmental appearance, sometimes shews on closer examination patches or bands with evident spherulitic growths or perlitic structure [604, 621, 605], or the much broken traces of convoluted fluxion-lines [604, 606]. Here the rocks seem to have suffered a certain amount of mechanical disturbance, and the network of perlitic cracks, as may well be imagined, has been specially favourable to the production of a false appearance of rounded fragments. Among the contemporaneous volcanic rocks of Westmorland, where also lavas have frequently been mistaken for ashes, the same passage from a perlitic rhyolite to a pseudo-fragmental rock is sometimes well exhibited [802].

It may be remarked here that many of the Caernarvonshire lavas shew signs of strain and sometimes of crushing. The most usual effect is seen in strings of granular secondary quartz traversing the slides. No doubt these often represent actual cracks filled in by infiltration or a leaching process, but in other cases they seem to be rather lines of maximum shearing strain than actual fractures. A felspar crystal lying across such a vein is slightly sheared but without disruption. This is well seen at Careg-y-defaid near Llanbedrog [700][1].

A rather different structure in some of the rhyolites may be referred to something of the nature of 'flow-brecciation' [678]. The rock is shattered into angular portions without displacement beyond the interposition of a network of veins. These veins consist mainly of quartz-mosaic, and may possibly be of secondary formation ; but they contain what appear to be glass-cavities, and seem to be not very sharply marked off from the original portions of the rock, which suggests a different explanation. In fact the structure may be of cognate origin with the quartzose lenticles described above and of the nature of 'contemporaneous' veining. We may suppose that a solid crust covering a still liquid though viscous flow was shattered by a network of cracks and the interstices injected almost at the same time with a highly acid 'mother-liquid,' which then consolidated in continuity with the rest of the rock.

Fragmental igneous rocks are known to exist in close as-

---

[1] Miss Raisin has noted a similar phenomenon, but offers a rather different explanation (*l. c.* p. 253 and fig. 4).

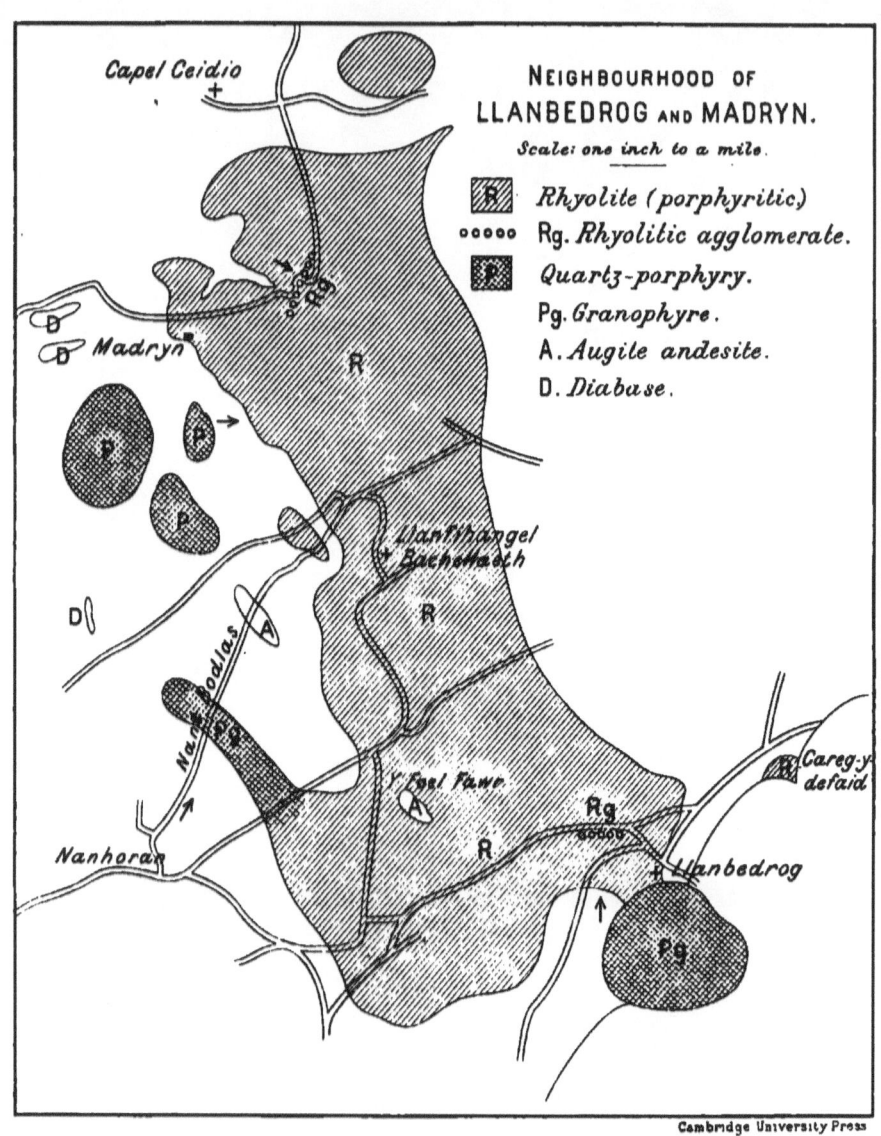

Fig. 2.

Capel Ceidio

NEIGHBOURHOOD OF
LLANBEDROG AND MADRYN.

Scale: one inch to a mile.

R. *Rhyolite (porphyritic.)*
Rg. *Rhyolitic agglomerate.*
P. *Quartz-porphyry.*
Pg. *Granophyre.*
A. *Augite andesite.*
D. *Diabase.*

Madryn

Llanfihangel
Bachellaeth

Nant Bodlas

Nanhoron

Y Foel Fawr

Careg-y
defaid

Llanbedrog

Cambridge University Press

sociation with the rhyolites at several places in south-western Caernarvonshire. The ashes seen close to Pwllheli are too much decomposed to allow of any description. In the Madryn and Llanbedrog district, farther west, there are one or two exposures of rocks, which are of interest as being made up of pieces of augitic granophyres, quartz-porphyries, etc., and so strengthening the evidence from other sources as to the community of origin of all the acid rocks in the district. The purplish agglomerate at the 'Pig Street' windmill near Llanbedrog shews, in a section [626], a multitude of crystals and rock-fragments united by a brown ferruginous paste. The crystals, not infrequently bent, consist of acid felspars of various kinds: the rock-fragments are all of quartz-porphyry and allied types. One of them is finely granular (micro-crystalline) with porphyritic felspars; another is a micro-pegmatite; and a third has a ground-mass of micro-pegmatite passing into spherulites, and contains green pseudomorphs with octagonal cross-section indicating augite crystals. This agglomerate occurs in the heart of the ordinary porphyritic rhyolites of this large igneous tract. (See map, fig. 2.)

Further to the north, at Y Gledrydd near Madryn, occurs a white or yellowish-white ash, associated with shales but doubtless connected also with the same period of vulcanicity which produced the rhyolites of the immediate neighbourhood. A slice of this rock [633] is seen to be composed of broken and bent crystals of quartz and felspars and portions of a spherulitic rock imbedded in a matrix of fine ashy material. The rock-fragments are chiefly detached spherules or portions of spherules. Mr Elsden records a similar rock from the south side of Mynydd Mynytho. The Llanbedrog and Y Gledrydd occurrences were noticed by Mr Tawney.

Both coarse and fine fragmental accumulations occur with the Pen-y-chain rhyolites, and have been described by Miss Raisin.

III. Nodular Rhyolites.

Among the various secondary changes to which the Caernarvonshire rhyolites have been subject, none is more striking than that evinced by the nodular rocks which we proceed to discuss. Under such names as '*roches globuleuses*,' '*pyromérides*,' 'globular porphyry,' 'ball rock,' 'concretionary felstone,' 'porphyry with agate nodules,' 'nodular felsite,' and 'coarsely spherulitic lava' these rocks have been described by numerous geologists, and more than one theory has been advanced to account for their peculiar structures[1]. Regarding all the Welsh varieties as different stages of one process of alteration, I shall consider them all together.

It is perhaps not generally known that these nodular structures are exceedingly common, occurring at all horizons among the acid lavas of Caernarvonshire. They are well exhibited near Yspytty Evan and Penmachno; at the Conwy Falls, Digoed, Glyn Lledr and other places near Bettws-y-coed; at Llangelynin, Y Ro-wen, and other localities south of Conwy, as well as at Conwy Mountain and Diganwy; in the hills above Llyn Crafnant and in the Penmachno Valley; at Llyn Ogwen, Llyn Idwal, and near Capel Curig; in Glyder-fawr, Esgair-felen, Cwmglas, and the head of Llanberis Pass; at Hafod-y-rhisg and other

---

[1] See Ramsay, pp. 122, etc. Delesse, *Mém. Soc. Geol. Franç.*, vol. IV., pp. 301—364; 1852. Bonney, *Q. J. G. S.*, vol. XXXVIII., pp. 289—296 and pl. X.; 1882. Rutley, "The Felsitic Lavas of England and Wales," *Mem. Geol. Surv.*, 1885. Cole, *Q. J. G. S.*, vol. XLI., pp. 162—168 and pl. iv.; 1885: vol. XLII., pp. 183—190 and pl. ix.; 1886. Miss Raisin, *Q. J. G. S.*, vol. XLV., pp. 247—269; 1889. Iddings, *Obsidian Cliff, Yellowstone National Park*, 7th ann. rep. U. S. Geol. Surv. 1888. See also numerous references in Messrs Cole and Iddings' papers, and description of Jersey nodules by de Lapparent, *Bull. Soc. Géol. Franç.* (3), vol. III., p. 223; 1875.

spots about Llyn Gwynant and thence to Beddgelert; in the country further south, on Moel-meirch, Yr Arddu, Mynydd Nant-y-mor, Moel-ddu, Oerddwr-isaf, and Moel-dinewyd, and in the pass between Beddgelert and Pont-Aberglaslyn. These localities include lavas belonging to all the different sets from the Upper Arenig flows near Llyn Conwy to the highest of the Snowdonian rhyolites. Similar varieties occur in the Lleyn promontory, not only in interbedded lavas, as at Pwllheli and Pen-y-chain, but in presumably intrusive rhyolites, such as those of Broom Hall near Afon-wen and Careg-y-defaid near Llanbedrog.

These nodular rhyolites may be compared with those of Skomer Island and Jersey, or still better with those of Westmorland, where on the line of the Bala lavas from Shap Wells to Stockdale magnificent examples may be collected.

The nodular structure usually occurs in the heart of masses of ordinary rhyolite, forming inconstant bands and patches which do not always correspond to the flow of the lavas. The nodules are often closely packed together. When best developed, they are very distinct from the enclosing matrix, and stand out by weather-ing, imparting to an exposed surface of the rock the appearance of a conglomerate of well-rounded pebbles. The size of the nodules usually ranges from one to three inches in diameter; but precisely the same phenomena are found on a scale varying from one-hundredth of an inch to as much as two feet. The large and the small nodules are often associated in the same rock, and their evident essential identity is important in discussing the manner of their origin. They are sometimes nearly spherical in form, but very often ovoid, and occasionally in the larger examples rather irregular, with some tendency to coalesce.

When nodules of the most usual kind are broken, they are found to present considerable differences, even in the case of speci-mens occurring side by side in the same locality. In some the eye perceives little or no difference between nodule and matrix. In others the outer part of the nodule resembles the rhyolitic matrix, while the interior consists of grey or yellowish flint. In addition there are frequently shell-like bands and irregularly shaped patches of a black substance. Some nodules, again, con-tain irregular hollows, not infrequently lined with mamillary chalcedony or with small quartz crystals coated with a yellowish-brown rust. Another type has the interior mainly occupied by greyish crystalline quartz. Finally there are nodules which possess no definite rounded outline, but consist of an irregular knot of white quartz, often sending out processes into the matrix, and enclosing a nucleus of the black material already mentioned.

A study of a large number of examples leads to the conclusion that these various types, with the probable exception of the last, represent successive stages of the most usual mode of origin of one original structure. The list might be indefinitely extended by taking account of minor modifications, such as those dependent on the fusion of several nodules or on the relative arrangement of the various secondary products. It has been remarked that a comparison of the larger with the smaller nodules is instructive with regard to their common origin. The processes which caused the differentiation of the nodules from the enclosing matrix have never advanced so far in the minute examples as in the generality of the larger specimens. The former rarely have distinct segregations of the black product so commonly met with in the latter. In the same rock we often find the larger nodules hollow or containing quartz, while the smaller have only flint; or the larger including patches of flint, while the smaller have not suffered any bodily replacement of their original substance. It should be noticed that the mode of occurrence of flint or crypto-crystalline silica in the nodules is such as to suggest a gradual molecular replacement of the original substance, while the quartz appears to have crystallised freely in previously formed cavities. The small nodules invariably give evidence of a well-marked spherulitic structure, and the least altered of the larger examples, when closely examined, shew the same character, though in a less evident degree. It is therefore natural to look to the spherulitic structure for the starting-point of the whole train of alterations embodied in the nodular varieties of rhyolite; and I believe that this is the right point of view for such examples, whether large or small, as have come under my notice.

Professor ·Bonney has put forward the theory that, in the. specimens examined by him, " the nodular or spheroidal structure has been produced in two ways:—(a) By simple contraction and roughly concentric cracking of the mass in cooling, being thus intermediate between the perlitic structure common in glassy acid lavas and the spheroidal structure common in basalt.... (b) By similar contraction in cooling, which is determined by the presence of a cavity...."

The sharply defined limits of the nodules of the ordinary types may indeed be regarded as in some sort a 'perlitic' structure, and the concentric zones and incomplete shells of black decomposition-products present in many varieties no doubt represent similar curved cracks widened by chemical actions; but these cracks seem to be different in origin from the true perlitic structure, which arises from simple contraction during cooling. Whenever the spherulitic structure is still to be detected, the fissures, if present, are seen to be clearly connected with it, forming a sharply marked boundary to the large or small spherules, and often a concentric system obviously related to that boundary. We must therefore seek analogies rather among the 'perlites' of Hungary, or in the spherulitic pitchstones of Spechtshausen in Saxony or the Lea Rock in Shropshire. Again, although actual cavities occur within some of the nodules, and others have hollows filled with crystalline quartz, the comparison of different nodules from any one good locality leads to the conclusion that these cavities were formed long subsequent to the consolidation of the rock, and owe their existence to the spherulitic structure still visible in the least altered types. The fact that the fluxion-lines traverse the nodules, but are interrupted by the cavities, may be taken to indicate very conclusively that the latter are not an original feature of the rock. Undoubted cases of infilled vesicles are met with, as already noted, among the Caernarvonshire rhyolites; but these present very different characters from the nodules here described, and are easily distinguished from them. The large size[1] of many of the true nodules would be sufficient to preclude the idea of their being of the nature of vesicles or 'lithophysæ,' and the structures described below are inexplicable except on the hypothesis that the nodules were originally solid throughout.

Most of these considerations have already been pointed out by Mr Grenville Cole, who has added much to our knowledge of these remarkable rock-structures. He lays stress in his papers on

[1] That this large size is not inconsistent with the spherulite theory appears from the dimensions attained by undoubted spherulitic growths in fresh Tertiary lavas. Cf. Whitman Cross, *Proc. Colorado Sci. Soc.*, 1887, pp. 230, 231.

the similarity between the minute spherulites so common in vitreous and quasi-vitreous acid rocks and the macro-structures for which he claims a similar origin. He compares the nodular rhyolites of Wales with various continental rocks, and particularly with the *pyromérides* of Bouley Bay in Jersey and Wuenheim in the Vosges. He further cites the opinion of Szabo, Roth, Zirkel, and Weiss, referring similar nodular structures in Hungary, the Yellowstone region, and the Thuringerwald to the alteration of large spherulites, as opposed to the "*Lithophysen*" theory of von Richthofen, which explains them as due to infiltration into original vesicles. The Caernarvonshire nodules differ from most of the European examples chiefly in the greater extent to which mineral replacement has proceeded, and the consequent further obliteration of the original structures; but this difficulty is removed on ex- amining in the field and in sections a series of specimens shewing different stages of alteration. Mr Cole describes in support of his view the large 'skeleton spherulites' of the Wuenheim rock[1], and his descriptions and figures present a very close resemblance to the less altered nodules at such places as Careg-y-defaid in the Lleyn peninsula. It is supposed that the structure of these large spherulites renders them more susceptible to destructive agents than the matrix in which they lie. In the case of the nodular rock of Digoed near Penmachno, our author traces the apparent process of alteration, beginning along minute cracks and ending in the total replacement of portions of the interior of the nodule. His analysis of the black substance occurring in these nodules is quoted below (I.). For comparison he gives the average (II.) of six analyses of the so-called 'pinite,' the composition of which varies, however, as is shewn in column III. There is, of course, no reason to suppose that the dark decomposition-product is the same in all the nodular rhyolites, or that it possesses any definite formula at all: its micro-structure presents many variations.

In connection with the alteration produced by weathering agencies, it is noticed by both Professor Bonney and Mr Cole that the matrix of the nodules often presents a slaty or schistose character. This observation, though by no means of general application, may perhaps have some significance. Professor Sedg- wick, whose Welsh note-books contain many references to these

---

[1] *G. M.*, 1887, p. 299.

	I.	II.	III.
SiO_2	50·75	47·8	44·7—54·6
Al_2O_3	28·34	29·9	23·6—32·4
Fe_2O_3	3·63	5·7	0·9—10·2
CaO	2·57	0·7	0·0— 2·4
MgO	1·85	1·8	0·0— 3·4
K_2O	6·21	8·7	6·5—11·2
Na_2O	2·19	0·5	0·0— 1·1
Loss	4·37	5·1	1·2— 7·8
	99·91	100·2	

'pseudo-conglomerates,' 'glandular concretionary rocks,' and 'agate-ball-rocks,' records an instance in which a nodular zone was found to coincide in direction with the cleavage of the rock, the 'bedding' having an opposite dip[1]. This was near the Conwy Falls, "just over the bridge leading to Penmachno." Such a phenomenon must be interpreted to mean, not that the spherulitic structure was developed along the cleavage-direction, but that the alteration of the spherulites was facilitated by the cleavage-structure.

Miss C. A. Raisin has recently made a careful study of the nodular rhyolites of Pen-y-chain and Careg-y-defaid, and classes the nodules in six groups :—

1. Contraction-spheroids or magnified perlitic structures.
2. Masses resulting from flow-brecciation.
3. Solid spherulites or pyromerides.
4. Agate-nodules, with an outer spherulitic crust.
5. Quartzose amygdaloids.
6. Spheroidal formations developed around a nucleus, such as an agate-nodule, a group of crystals, or an original vesicle of the lava.

The evident amygdaloids or infilled vesicles (5) may be dismissed, being easily distinguished from the nodules proper, and their origin undisputed[2]. Classes 4 and 6 represent in my view, as in

[1] MS. Note-book No. XXII., in the Woodwardian Museum.
[2] See figs. 8 and 9 in Miss Raisin's paper.

Mr Cole's, various stages of alteration of Class 3, the original spherulite being destroyed in the interior, which is replaced by new products, or an actual cavity formed which has been filled again by chalcedonic silica or crystallised quartz. As regards Class 1, I have not met with nodules in the rhyolites which appeared to be due to simple contraction from cooling, apart from any spherulitic growth: the phenomena seem rather to point to concentric cracks being produced around a spherulitic centre, owing to molecular changes within the domain of the spherulite. Miss Raisin's suggestion of silicification by geyser-action is one on which it is not easy to form an opinion.

It seems probable that the comparison of our nodular rhyolites with the Obsidian Cliff rock in Yellowstone Park is a misleading one. The description of this rock by Mr J. P. Iddings and the admirable figures accompanying his memoir shew very considerable differences from the Welsh nodular rhyolites, and make it appear that the remarkable chambered structures in the American obsidian are in some sense true lithophysæ. Although Mr Iddings' explanation of them differs in some respects from von Richthofen's, he still considers the cavities to be due to the action of water-vapour. The connection between the spherulitic growth and the release of the previously absorbed vapour he expresses thus: " In the still viscous glass, from a centre of crystallisation the first frail beginnings of felspar spread in innumerable rays, pre-empting, as it were, a sphere of the magma. The enlargement of these anhydrous microlites by crystal growth from their matrix of hydrated glass...rendered it relatively more hydrous, so that, with decreasing temperature, it may no longer have been able to retain the vapours in combination." " Upon the change of the hydrous paste to a crystalline aggregate through the process above indicated there would be a considerable shrinkage, the extent of which is indicated by the gaping cracks and parted segments characteristic of lithophysæ." The author does not ascribe the production of the cavities in any degree to the expansion of the released vapour, nor regard the lithophysæ as due to the existence of any original vesicles in the rock.

Although the theory thus briefly sketched seems to account very satisfactorily for the phenomena of the lithophysæ at Obsidian Cliff, it would be rash to apply it to the case of the

Caernarvonshire nodules, where the relation of the cavities to the enveloping shell is so different. What I have been able to observe compels me to embrace Mr Cole's theory of the derivation of all the true nodules from simple spherulites, rather than any of the explanations offered by Professor Bonney and Miss Raisin postulating an original cavity. The examples studied in the field and in specimens include the rocks described by the three writers just named, as well as others from most of the Welsh localities named at the beginning of this section, and even finer specimens from Westmorland. I am indebted to the Rev. E. Hill for the specimens from Jersey which I have also examined.

In view of the endless variety of detailed structure found among the nodules, and of the discussion they have already received at other hands, I shall confine myself here to describing, with but little comment, a few of the more striking types.

No locality affords more varied and instructive examples than the little abrupt rock named Careg-y-defaid, between Pwllheli and Llanbedrog, noticed on the Survey map as 'felspar-porphyry with agate nodules'. It is a porphyritic rhyolite, exhibiting evident flow-structure on a small scale, and in some parts, as is well seen on the sea-shore, it is crowded with nodules of all sizes up to two feet or more in diameter. Occasionally the larger ones are seen to be set in a mass which itself consists to a great extent of little nodules about the size of peas. As usual, the larger specimens are of less regular forms than the smaller. Taking those of moderate dimensions—say two or three inches across—it is seen that some, when broken open, present no sensible difference from the ordinary rhyolite[1], and the same is true, at this locality, of the outer crust of all the spheroids. On microscopic examination, however, [701] a certain difference is perceived between the rhyolite within and without the well-defined boundary of the nodule. The former, like the latter, includes porphyritic felspar crystals and is traversed by traces of the fluidal lines; but quite independent of these lines of flow, and for the most part obliterating them, there is seen also a faintly defined radial

---

[1] The features here described are in close agreement with those recorded by Whitman Cross in the Tertiary rhyolites of Custer Co., Colorado, *Proc. Col. Sci. Soc.*, 1887, pp. 230, 231.

structure, which can be regarded only as a giant 'skeleton-spherulite'.

The specimen sliced does not shew the central part of the spherulite, this having been replaced by a kernel of amorphous silica. The latter is of irregular form, and the manner in which it protrudes in places into the outer part of the nodule, interrupting the lines of flow, which shew no sign of bending around it, is conclusive as to its occupying the place of decomposed rhyolitic substance and not merely filling an original cavity. It seems probable that this amorphous silica, which I have referred to as flint, does not even fill cavities produced by the weathering out of the internal portion of spherulites, but is rather the result of gradual molecular displacement. In this respect it seems to differ from the quartz. The latter mineral is found *partially* filling cavities in the nodules of various localities, but this does not appear to be the case with the flint, although chalcedonic silica in a mammilary coating sometimes lines cavities in the Careg-y-defaid nodules. The dark decomposition-product in these nodules occurs partly in inconstant shell-like layers, as if filling widened shrinkage-cracks, and partly in shapeless nuclei. These shrinkage-cracks, as well as the more regular one which commonly bounds each nodule, I would refer to condensation accompanying the incipient crystallisation of the spherulite, rather than to contraction from cooling as in ordinary perlitic structure. That the fissures were originally very narrow and have since been enlarged by chemical action can sometimes be verified by the behaviour of the fluxion-lines when interrupted by the cracks. The appearances are quite different from those observed in connection with the chambers in Mr Iddings' lithophysæ.

A specimen from Conwy Mountain has egg-shaped nodules about an inch and a half in length, consisting chiefly of rhyolite which to the eye resembles that forming the matrix. Lines of flow marked by streams of magnetite granules pass through the nodule without deviation. A polished section shews, however, a faintly indicated radial structure in the external part of the nodule, and this is crossed by a much more pronounced concentric structure parallel to the outer boundary. The centre is occupied by an almond-shaped kernel of the usual flinty material.

The obliteration of the radial spherulitic structure by a more marked concentric structure characterises one common kind of alteration of the original spherulite. The process appears to consist in the bringing to light of minute concentric cracks, which become widened and occupied by the dark decomposition-products. The shells of rhyolite separating these cracks may become silicified as a further change, but some of the nodules with this concentric structure shew but little of the flinty replacement. Many of the Diganwy specimens present a succession of black ellipsoidal shells, one within another, and Mr Cole has figured something similar from Conwy Mountain.

The independence of the concentric fissures in nodules and the true perlitic structure in the rhyolitic mass is well illustrated in some specimens from Pwllheli. The nodules form ellipsoids up to three or four inches in length, and in thin sections [662] exhibit admirably the manner in which the patches of flint interrupt the fluxion-lines plainly visible in the outer crust of the nodules. This outer crust, which is in the main of ordinary rhyolite, is traversed in places by a most delicate system of curving perlitic cracks; and the amorphous flint is everywhere seen bordering these cracks and eating its way into the rhyolite between them. In other places the shreds of flint are observed to follow the lines of flow, with a tendency to develope in the bends of curves. We may gather from these specimens that the flinty material is formed rather by replacement of the original substance than by the infilling of cavities, but that this process of silicification is facilitated by cracks or other surfaces of discontinuity or weakness, and may then constitute rather an early stage in the transformation of the original spherulites into nodules.

An interesting specimen from Llyn Ogwen contains small and comparatively large ovoid nodules in one slice [661], and illustrates very well the more rapid advance of alteration in the larger individuals. The smaller ones, some not more than a hundredth of an inch long, are entirely occupied by amorphous silica, and still retain in the flinty pseudomorphs traces of an original radial structure. The larger ones have an outer flinty portion, while the interior is occupied by quartz. Not only has the original rhyolitic substance entirely disappeared, but all trace of the structure of the spherulite is lost, except its ellipsoidal boundary.

The presence of quartz in the heart of the nodule always seems to indicate an advanced stage of alteration. In the specimens just described this mineral occurs in a very finely granular form, and it is probable that no considerable hollow has ever existed for free crystallisation to take effect. In most cases, however, the mode of occurrence of the quartz can be explained only by infiltration into a central cavity, often of considerable relative dimensions, which we must suppose to have arisen from the destruction of the inner portion of the spherulite after the formation of the external flinty crust. In many examples the quartz occupies the greater portion of the total volume of the nodule.

A specimen from the Penmachno valley [659] has the nodules largely composed of quartz disposed in a coarsely granular mosaic. The mineral is traversed by lines of minute fluid-cavities, arranged in two sets, and the lines pass uninterrupted through the several grains of the mosaic, as figured by Mr Cole in the Digoed rock from the same district.

The types of nodules briefly described above, and many modifications of them, are linked to one another by intermediate varieties, as may be seen on studying them in the field, many transitions being sometimes seen at one locality; so that it is impossible to avoid grouping them all together. I have found exceedingly few nodular structures among the Caernarvonshire rhyolites, which leave any room for hesitation as to whether they should or should not be referred to altered spherulites. One doubtful case is that of a rock from the neighbourhood of Llyn Crafnant, in which occur quartz-nodules of irregular outline. They appear not to be sharply bounded, but to send out processes and veins, or even to pass gradually, by a finer granulation of the quartz-mosaic, into the rhyolitic matrix. A section [660] shews that the outer portion of the nodule consists of well-formed quartz crystals pointing inwards, while the interior is filled with a coarse quartz-mosaic, excepting a central kernel of black material. This last is seen under the microscope to consist of bundles of a bright green mineral with vermicular form, and it is included by the quartz-mosaic in such a way as to prove its earlier formation. Each of the worm-like bodies consists of a *rouleau* of little discs, apparently of hexagonal outline, piled one on another, and the mineral appears to be one of the chlorite family. Professor Bonney records from a rhyolite near the Conwy Falls a vermicular chloritoid substance, which he considers to agree generally with delessite, but our mineral, which gives fairly high double refraction, is more like the chlorites proper.

If these nodules are to be classed with the rest, they would seem to represent the most advanced stage of alteration, in which even the boundary of the parent spherulite is effaced. It is noticeable, however, that the fluxion-lines in the matrix, though broken by the veins and processes of quartz, seem on the whole to wind round the nodule; and it is quite possible that this structure has its origin in a true vesicle complicated by later fissures. The knots of quartz are never more than an inch long.

In describing a few examples of the ordinary nodular rhyolites of the Caernarvonshire area, I have noted chiefly those points which seem to elucidate the origin of the structures in question. There is still much that would repay a detailed study by any geologist interested in the secondary transformations of igneous rocks. The general conclusion is that all the true spheroidal nodules owe their origin to solid spherulites of very various dimensions. In most of these spherulites the radial structure was well marked, and the centre of the spheroid, where the delicate radiating fibres all terminated, was the part most susceptible to the operation of destructive agencies, so that the replacement of the rhyolitic substance by amorphous silica has usually begun there. The slight shrinkage consequent upon the molecular rearrangement in the spherulite, after the enveloping rock had become effectively solid, determined a sharp boundary between the spherulite and its matrix, and sometimes concentric spheroidal fissures within the boundary, while irregular cracks appear to have been formed in many instances from the same or from more general causes. When this was the case, the process of alteration began in a different way, frequently proceeding in concentric shells. In all cases where the destruction has sufficiently advanced, the interior of the nodule seems to have been completely removed, with the exception of the pinite-like substance, and a cavity was formed, which, however, is usually filled, completely or in part, by crystalline quartz.   This removal of the interior of the nodule has taken place even through a previously formed shell of flint.   The chemistry of this process has been discussed by Mr Cole, the disappearance of the alumina being the chief difficulty to be explained.

THE rocks next to be described, though acid in constitution, differ both geologically and lithologically from those discussed in the preceding pages. In the first place, they are all intrusive, while the great majority of the rhyolites are extrusive; and secondly, instead of the volcanic *habitus* of the lavas, they present the petrological types characteristic of bosses, necks, and laccolites of acid rocks. They may be described comprehensively as biotite- and especially augite-bearing granite-porphyries, granophyres, and quartz-porphyries. These rocks shew, however, abundant evidence of belonging to the age of the Bala volcanic series and being related in a definite manner to the rhyolitic lavas. We shall see, on the one hand, that the acid irruptives, more particularly in eastern Caernarvonshire, mark the sources from which the lavas were poured out during the Bala age; and, on the other hand, we shall have occasion to notice in the south-western part of the county tracts shewing every gradation of petrological character between the characteristic type of the larger bosses and that of the Snowdonian rhyolites. All the intrusions occur nearly on a line drawn south-westerly from Y Foel Frâs to Carn Fadryn, and it will be convenient first to describe in order their positions and probable mode of occurrence.

To the south of Aber and Llanfairfechan a tract of igneous rocks occupies the high ground which culminates in Y Drosgl to the west and in the rounded summit of Y Foel Frâs, some 2500 feet above sea-level, to the east. The rocks crop out in an irregular oval, three and a half miles in diameter, broken off on the north side by the Aber and Llanbedr fault. · Sir A. Ramsay

(p. 139) describes them as sometimes felspathic porphyry, sometimes hornblendic greenstone, and states that these varieties, separately coloured on the Survey map, "pass gradually into each other." This view is not sustained by what I have seen of the rocks, and the 'greenstone' is partly diabase, partly andesite and andesitic ashes. Excluding these for the present, the great bulk of the igneous area consists of augitic granophyre. Its intrusive nature is attested by the contact-alteration of the adjacent strata at Aber Cascade and other places, but the modification is not very extensive. The behaviour of the cleavage in the slates of the vicinity is consistent with the supposition that the rocks rise through the neighbouring strata in the form of a large boss with roughly vertical walls.

A mile or two from this large mass a little intrusive patch is mapped as greenstone, but it is stated (Ramsay, p. 172) that it "sometimes approaches syenite in its structure", and on the south side of Cwm Gaseg sends veins into the slates. This small mass situated on the north-west slopes of Yr Elen, I have not been able to visit, but it will probably be found to be only a diabasic intrusion and so to have no place in this part of our subject.

The intrusive mass exposed on Moel Perfedd, a hill overlooking Nant Ffrancon, has also been noticed by Sir A. Ramsay (p. 172 and fig. 60), who gives a figure to explain its relation to the neighbouring strata. It is clearly a partly denuded laccolite, injected wedge-like between grits and flags of Lingula age. It visibly metamorphoses the adjacent strata on both sides.

I shall venture to describe as laccolites intrusive masses whose relations are less clearly laid bare than those of the Moel Perfedd rock, and a few words on this subject will therefore not be out of place. The idea of any extensive incorporation or 'eating up' of sedimentary strata by an invading igneous magma is negatived by all observations of the junctions of such rocks in this country. This being so, the laccolite remains as the normal and natural form which must be assumed by any large intruded mass which does not cause violent disturbance of the adjacent strata. When the intrusion is effected among rocks not modified at the time by superinduced structures, the extension of the laccolite will naturally be along the bedding-planes, which are in such a case the surfaces of least cohesion. If the country be subsequently dis-

turbed, and the rocks become inclined and suffer partial denudation, it will usually be very difficult to establish directly the
laccolitic nature of the intruded body of rock; but the characteristic form of the outcrop will be an elongated oval or lenticular
patch on the map, having its long axis in the direction of strike of
the neighbouring strata. The point which is generally overlooked
is that this elongated form of outcrop and parallelism with the
local strike are in themselves the strongest circumstantial evidence
that the intrusion is a true laccolite, and indeed admit of no other
general explanation. There is no conceivable reason why an
igneous mass should have its greatest extent of outcrop in the
direction of strike, unless the mass be intruded along bedding-
planes. Such a mass may, of course, without inconsistency break
through the strata in some places, and a laccolite in direct connection with a small transgressive boss of similar rock may be
expected as a not unusual phenomenon.

An illustration of this last remark is afforded by the next
intrusion to be noted, which we shall name that of Bwlch-Cywion.
It lies about a mile to the south-east of the preceding, and
is of similar lithological type, though less granophyric. It appears to be clearly transgressive at the end nearest Nant Ffrancon,
and sends out veins into the neighbouring slates, which become
porcellanised at the junction[1]; but it extends in a south-westerly
direction in the form of a shallow laccolite.

The next in the chain of intrusions is the mass which forms
the rounded hill Mynydd Mawr, rising to a height of 2300 feet
immediately west of Llyn Cwellyn. Notwithstanding Dr Hicks'[2]
claim that this is part of a pre-Cambrian land-surface, the intrusive
nature of the boss is sufficiently evident from the intense alteration it has produced in the slates in contact with it, probably of
Arenig age. From the manner in which this rock has modified
the direction of cleavage in the slates, I have shewn elsewhere[3]
that it is a cylindrical neck, and is of age anterior to the maximum
crust-movements in the district. This fixes the time of intrusion
as within the Bala period, and the line of argument exemplified
in the paper cited is applicable to other cases. Petrologically the

[1] Ramsay, p. 172.
[2] Q. J. G. S., vol. xxxv., p. 297; 1879.
[3] G. M., 1888, pp. 221—226.

Mynydd Mawr rock, though clearly allied to the rest, has sufficient interest to be worthy of special notice.

The peak of Y Garn, two miles south of Llyn Cwellyn, though mapped as greenstone, is stated by Sir A. Ramsay (p. 171) to consist of a rock similar to that of Mynydd Mawr, while the slates on its western side dip beneath it. If this be verified, the mass may possibly be a laccolite connected with the Mynydd Mawr neck.

It may be mentioned, to avoid needless prolixity, that all the remaining rocks have also been claimed by Dr Hicks[1] as pre-Cambrian; but in every case I have found manifest contact-metamorphism and abundant other evidence that the rocks in question are intruded[2] among the neighbouring strata, which range from the Arenig to the middle Bala. A number of speci-mens of these rocks have been described by Professor Bonney[3] and Mr Tawney[4], but their relations to one another have hitherto received little attention.

Proceeding south-westerly, we come to the Clynog-fawr district, where a number of distinct intrusions occur, which unfortunately have not all been studied in detail. There is a chain of four large hills due to a set of connected intrusions, while in the hilly country immediately south occur four isolated masses. All these are acid rocks, although one has been erroneously mapped as 'green-stone'. The specimens from the northerly chain are all augitic granite-porphyries and transitional types allied to this. There may be a granophyric tendency, as in the Girn-ddu quarries; or gradations in the ground mass into a more finely crystalline variety, as at the top of the same hill; or the quartz-porphyry type may prevail, as in Girn-goch. In the most easterly hill, named Bwlch-mawr[5], there are probably several gradations of texture and structure. A specimen from the top, 1678 feet above sea-level, is a typical augite-bearing granite-porphyry, while an-

---

[1] Q. J. G. S., vol. xxxv., pp. 297—299; 1879.

[2] Cf. Blake, Q. J. G. S., vol. xliv., pp. 533, 534; 1888. Sedgwick, ibid., vol. iii. p. 148; 1847.

[3] Ibid., vol. xxxv., p. 305. G. M., 1880, p. 458.

[4] G. M., 1882, pp. 552, 553; 1883, pp. 17—21.

[5] This name, signifying 'the great pass', probably applies to the deep hollow on the south side of the hill.

other, taken lower down on the north-side, presents a close resemblance to the Penmaenbach rhyolites. The other intrusions of the neighbourhood are quartz-porphyries of various characters, the Pen-y-gaer rock being a beautiful granophyre, while the little intrusive bosses of Cil-y-coed north-east of Clynog-fawr and Moelfre near Llanaelhaiarn shew rhyolitic affinities, and the same is probably true of Mynydd-y-cennin the most easterly intrusion of this neighbourhood (Ramsay, p. 218).

Next we arrive at the district of the seven hills popularly known as the Rivals, where the map shews two small and one large patch of igneous rocks. The two smaller are of augitic granite-porphyry more or less granophyric: they are extensively quarried, especially at Trefor on the sea-ward side of the northerly mass. The large area stretching to the south shews quartz-porphyries (also augitic) varying from granular and granophyric varieties about Yr Eifl and Tre'ir Ceiri to a thoroughly rhyolitic type in the south, as seen near Carnguwch and Melin Llwyn-dyrus.

Augite-bearing granite-porphyries and granophyres occur again in force in the Nevin district, being quarried near the town itself. Sometimes, as at Moel-gwyn near Pistyll, a less acid type is found, and this is perhaps better described as syenite-porphyry; but there can be no doubt that all these igneous rocks in the northern part of the tract are closely related to one another. An examination of the rocks to the south would perhaps discover gradations into a more volcanic type. The Carn Boduan rock is quite distinct from the rest, and will be described in its place among the andesites.

The fine hill named Carn Fadryn (1221 feet), with its smaller companion Carn Fach, owes its prominence, as shewn on the Survey map, to three intrusive masses of augite-bearing quartz-porphyry with occasional granophyric tendency and various points in common with the other acid irruptives of the county.

Other quartz-porphyries, chiefly granophyres, occur not far from the last, and are evidently related to the foregoing, while closely associated with the rhyolites of the district. Of this kind are the tongue, probably a thick sheet intruded along the bedding, which crosses Nant Bodlas a mile above Nanhoran, and the oval boss of Mynydd-tir-y-cwmmwd south of Llanbedrog (see map, fig. 2). The rest of the tract between Madryn and Llanbedrog is occupied, as already stated, by porphyritic rhyolites with some rhyolitic agglomerates, broken through at Y Foel Fawr and possibly in other places by andesites. A fine 'pseudo-spherulitic' quartz-porphyry occurs in a little boss immediately south of Penmaen Castle near Pwllheli (see map, fig. 3), and the wide distribution of rocks of this type is proved by their occurrence in the fragmental volcanic accumulations at various localities, as already noted.

A general idea of the chemical composition of the rocks in question may be obtained from the following analysis (I.) of a granophyric granite-porphyry from Y Drosgl, a hill forming part of the Y Foel Frâs tract. The specific gravity of the specimen was found to be 2·772.

	I.	II.	III.
SiO_2	70·6	71·442	72·79
Al_2O_3	13·3	15·340	13·77
Fe_2O_3	3·1	1·230	3·32
FeO	not est.	1·107	—
MnO	—	—	trace
CaO	2·2	1·064	1·94
MgO	0·4	0·720	0·62
K_2O	9·2	4·439	2·99
Na_2O	0·8	3·951	4·12
P_2O_5	—	0·118	—
SO_3	—	trace	—
CO_2	—	trace	—
Loss on ignition	0·3	0·589	1·08
	99·9	100·000	100·63

I. Granophyre, Y Drosgl near Aber. Anal. E. Hamilton Acton. (All the iron is estimated as ferric oxide.)

II. Granophyre, Scale Force, Buttermere. Anal. J. Hughes; Q. J. G. S., vol. xxxii., p. 22; 1876.

III. Granophyre, Llyn-y-gader, Cader Idris. Anal. T. H. Holland; Q. J. G. S., vol. xlv., p. 435; 1889.

The figures indicate about 54 *per cent.* of potash-felspar and perhaps half that quantity of quartz. On comparison with the

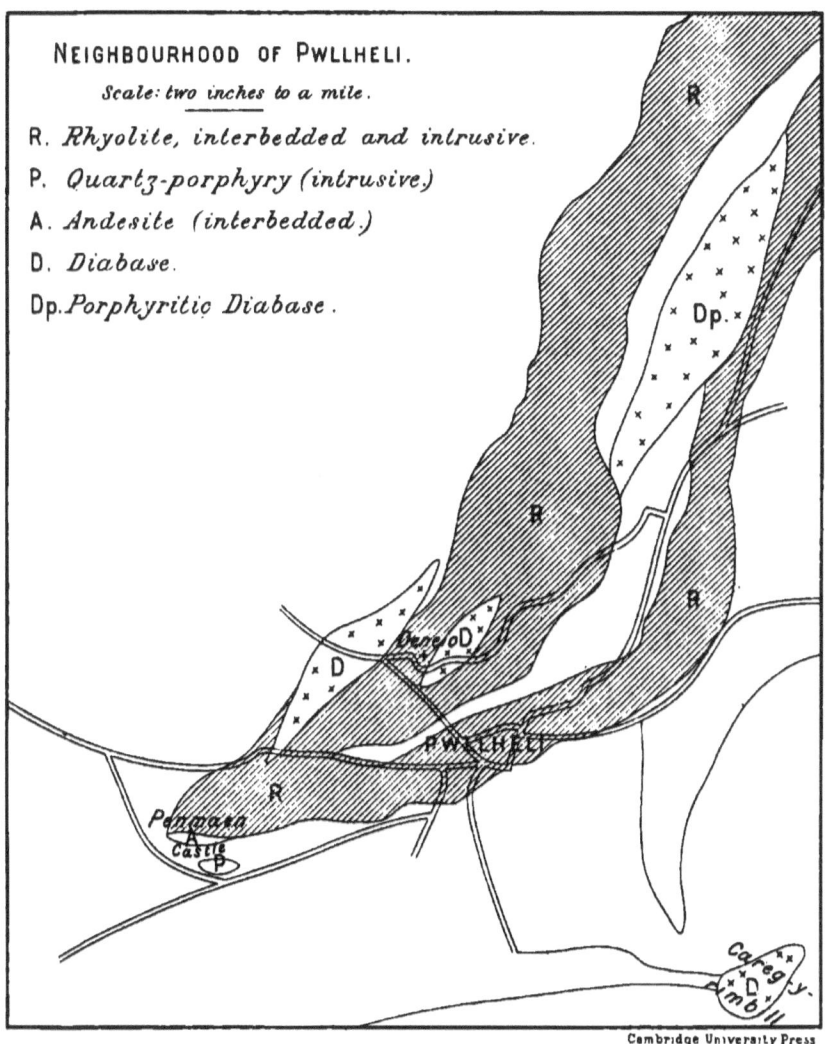

NEIGHBOURHOOD OF PWLLHELI.

Scale: two inches to a mile.

R. Rhyolite, interbedded and intrusive.

P. Quartz-porphyry (intrusive.)

A. Andesite (interbedded.)

D. Diabase.

Dp. Porphyritic Diabase.

Cambridge University Press

Fig.3.

analyses of acid lavas from Caernarvonshire given on a former page, it will be seen that the granophyre is less acid than the least altered rhyolites. Like them it shews an unusually high percentage of potash. In the relative proportions of the alkalies it differs notably from the Lake District granophyres, which are in other respects closely similar to those of Caernarvonshire, and doubtless stand in a similar relation to the Bala rhyolites of the district. An analysis (II.) of the Buttermere granophyre ('syenitic granite' of Mr Clifton Ward) is given for comparison. The Cader Idris granophyre ('eurite' of Messrs Cole and Jennings) has soda in excess of potash (III.).

The Mynydd Mawr rock yielded, according to Mr Acton, about 10 *per cent.* of potash and 4 *per cent.* of soda. These unusually high percentages of the alkalis shew themselves in the remarkable alkali-bearing amphibole, riebeckite.

Hand-specimens of the rocks, as might be expected, shew much variation, and indeed the close relationship which runs through the whole series is only to be seen in the microscopical sections. The granite-porphyries and coarser granophyres have the appearance of fine-grained granites, in which occur porphyritic felspars, sometimes scarcely larger than the grains of the groundmass (Nevin), sometimes an inch or more in length (Bwlch Mawr and Y Girn-ddu). When weathered the felspars are opaque white, when fresh they are almost glassy and shew Carlsbad or albite twinning. The ground-mass is greyish-white, rarely reddish, and speckled or dotted with black. The augite is never clearly identifiable, but the biotite can often be seen with a lens (Trefor and Nant Gwrtheyrn), and in rare cases builds isolated flakes a quarter of an inch in diameter. Little flakes of white mica are found in the Nant Bodlas rock only. The black patches, from two to four or five inches across, which are common in some of the rocks, will be noted below. In the more minutely granophyric varieties of rocks and in the quartz-porphyries these patches are not seen. The ground-mass of these latter rocks is dull grey or yellowish and exceedingly compact in appearance, and the porphyritic elements, often including quartz as well as felspar, are usually quite small.

We proceed to the microscopical examination[1], first treating

---

[1] More than forty slides have been studied.

all the varieties as far as possible together, and reserving their differences until the discussion of the nature of the ground-mass. The sections are found to consist of a ground-mass varying from granular to granophyric, in which are embedded porphyritic crystals of felspar, commonly from a tenth to a quarter of an inch in length.

Distinctly porphyritic quartz is developed, as a rule, only in the varieties here designated quartz-porphyries, where it forms rounded grains or pyramidal crystals. Where quartz occurs among the earlier constituents in the granite-porphyries, as in the Bwlch Cywion mass [316, 317], it is in rounded grains enclosed by the porphyritic felspars, and contains fluid cavities with spontaneously moving bubbles.

The porphyritic felspars are chiefly plagioclase of the albite-oligoclase series, in which the albite-lamellation is often combined with twinning on the Carlsbad or the pericline law or both. The crystals are well-bounded and present rectangular sections. Ortho-clase occurs more sparingly, and is of slightly posterior formation, sometimes moulding the triclinic felspars. Squarish sections exhibiting minute and rather shadowy cross-hatching in polarised light seem to be microcline : they occur only in some of the typical granite-porphyries, such as that of Bwlch-Cywion [316, 685]. The plagioclase crystals often display a zonary banding in polarised light, due to a variation in chemical composition from the inner to the outer layers[1]. The extinction-angles seem to indicate that a felspar near oligoclase in composition is bordered by albite. This is seen in specimens from Trefor (Yr Eifl) and other places [40, 670, etc.]. More remarkable is the bordering of the lamellated acid plagioclase by a margin, 0·001 to 0·005 inch wide, of uustriated felspar with the appearance of orthoclase. This is seen at Trefor, Y Drosgl (three miles south of Aber), Moel Perfedd, Moel Penllechog, and Y Girn-ddu [40, 83, 628, 702, 693, 738, etc.]. A zonary banding is sometimes observable in crystals which shew no twinning, and is perhaps to be explained by a varying iso-morphous mixture of potash- and soda-orthoclase.

Apatite in colourless needles is seen in several slides from

[1] Applying M. Michel-Lévy's test, it does not appear that the zoning can be explained in general by ultra-microscopic twinning: the zones do not always disappear with the albite-lamellæ on rotating the sections beween crossed nicols. Cf. *Minéraux des Roches*, p. 85; 1888.

Trefor and the Nevin district, and the augitic quartz-porphyry of Carn Fadryn has the same mineral in larger hexagonal prisms with rounded inclusions.

Magnetite forms ragged grains in many of the specimens, but rarely takes crystal forms, except in some of the segregation-nodules of the granite-porphyries, where it is very abundant, and where it occurs included in the porphyritic felspars. This mineral and the apatite are of earlier consolidation than the other constituents of the ground, and often earlier in part than the porphyritic felspars. Light brown grains of sphene are very rarely met with [634].

A highly characteristic feature of all the rocks is the abundant presence of augite and biotite, to the exclusion of original hornblende. Biotite is found in all the examples from the eastern division, as well as at Trefor, while augite occurs in all the rocks with only the exceptions of Mynydd Mawr and Bwlch Cywion, where its place is taken by a remarkable amphibole-mineral. The biotite is normally of the brown variety, but often turns green, giving then a rather pale grass-green tint for vibrations parallel to the cleavage-traces and a paler yellowish-green or greenish-yellow for the perpendicular direction. The mineral is usually confined to the ground-mass, where it precedes the quartz and felspar in the order of consolidation.

The augite builds idiomorphic crystals, but often rather rounded in outline, and the numerous granules in the segregation-nodules are quite round [671, etc.]. When fresh, the mineral is colourless in thin sections, and shews the cleavage in well-marked fine cracks. A peripheral change into green or brownish-green hornblende is a very common feature, although in some examples, such as those from the Nevin district, it is wanting. The common weathering processes give rise to the familiar pale-green chloritoid minerals, occasionally with radial grouping about centres which give a black cross between crossed nicols [16, 686]. Epidote is also present in a few slides [16, 693], but in most cases this product is rather connected with the alteration of the felspars [739, etc.]. The augite is an early product of consolidation, being frequently enclosed by the first generation of felspar.

The only mineral of the amphibole family found as an original constituent of any of the acid rocks of Caernarvonshire is the singular blue variety which occurs in the Mynydd Mawr rock, and possibly elsewhere. The rock in question appears in hand-specimens as a fine-grained grey mass, weathering cream-coloured or pink, enclosing lustrous black crystals arranged in fluxional streams. These crystals range up to about a quarter of an inch in length, and have an irregular outline. A few small felspars are also seen, but only occasionally a grain of quartz.

The black crystals are the hornblende of Sir A. Ramsay's description (p. 171), but they belong to a variety not elsewhere dis-covered in Britain. In recording the presence in the Mynydd Mawr rock of a new variety of hornblende, I was at first deceived by its exceptional properties and mistook a part of it for tourmaline[1]. The mineral was independently discovered by Professor Bonney, who described it as arfvedsonite[2]. From this, however, it differs con-siderably, and there can now be no question as to its identity with the *riebeckite* of Sauer[3], which occurs in a granite from Socotra. The presence of this mineral in two localities so distant is not a little curious. In this connection, too, it would be interesting to have a more detailed description of the 'intensiv blaugrün gefärbte Hornblende' from the island of Sikoku (Japan) mentioned by Oebbeke[4] and Koto[5] as distinct from glaucophane and shewing much more intense coloration and pleochroism[6].

In the Mynydd Mawr rock, as in that from Socotra, the rie-beckite occurs in two very different habits; *viz.* in the crystals already mentioned and in mere microlites. The crystals have an exceedingly ragged appearance owing to their enclosing numerous grains of the ground-mass, both at the border and, to some extent, in the interior. This indicates that they were not clearly separated at an early stage of the consolidation: the crystals occasionally mould the squarish porphyritic felspars. Slices cut thin enough to be transparent shew the prismatic hornblende-cleavage. Rie-

---

[1] *G. M.*, 1888, pp. 221, 455.

[2] *Min. Mag.*, vol. VIII., p. 103; 1888.

[3] *Zeitschr. d. deutsch. geol. Gesellsch.*, vol. XL., p. 138; 1888.

[4] *Zeitschr. d. deutsch. geol. Gesellsch.*, vol. XXXVIII., pp. 641, 653; 1886: *Zeitschr. für Kryst.*, vol. XII., p. 285; 1886.

[5] *Journ. Coll. Sci. Tokio*, vol. I., p. 303; 1887.

[6] Since the above was written, M. Urbain Le Verrier has announced the discovery of riebeckite in a granulite from Corsica: *Paris Acad. des Sciences*, July 1st, 1889.

beckite differs optically from all the other amphiboles. As was pointed out by Professor Rosenbusch[1], the axis of elasticity which makes a small angle with the vertical axis of crystallography (c) is not γ but α. The pleochroism is very strongly marked: thus—

$\alpha \gtrless \beta >> \gamma$;

α, very intense indigo-blue;

β, a slightly less intense blue, with greenish tinge;

γ, rather light brown or greenish-brown.

The absorption for α and β is such as to render any but very thin slices opaque. No other known mineral offers the same characters. The extinction-angle ($c\alpha$) in clinopinacoidal sections is less than 5°.

No chemical analysis has been made of the Welsh riebeckite, but I quote below Dr Sauer's analysis of the Socotra mineral (I.). The other columns, given for comparison, are Lorenzen's analysis of ægirine from Greenland (II.) and Dölter's analysis of a Greenland mineral (III.), which he described as arfvedsonite, but which, as

	I.	II.	III.
SiO_2	50·01	52·22	49·04
Al_2O_3	—	0·64	1·80
Fe_2O_3	28·30	28·15	29·54
FeO	9·87	5·35	4·82
MnO	0·63	0·54·	trace
CaO	1·32	2·19	2·70
MgO	0·34	1·45	trace
K_2O	0·72	0·34	trace
Na_2O	8·79	10·11	13·31
Loss by ignition	—	—	—
	99·98	100·99	101·21

I.   Riebeckite, Socotra.   Anal., A. Sauer, *l.c.*

II.   Ægirine, Greenland.   Anal., J. Lorenzen; *Meddelelser om Grönland*, Part II., p. 55; 1881.

III.   Ægirine (so-called Arfvedsonite), Greenland. Anal., C. Dölter; *Zeitschr. für Kryst.*, vol. IV., p. 34; 1880.

[1] See Sauer's paper *cit. sup.*

pointed out by Lorenzen, was probably ægirine. The figures are sufficient to prove the substantial identity of chemical composition of riebeckite and ægirine. Indeed, as Professor Rosenbusch remarks, the former mineral holds precisely the same place among the amphiboles as the latter does among the pyroxenes, the resemblance between the two being borne out by the position of the axes of optical elasticity, which differentiates riebeckite, ægirine, and acmite from all other members of the amphibole and pyroxene families. From his analysis of the Socotra riebeckite Dr Sauer deduces the formula

$$\begin{cases} 5 \ FeSiO_3 \\ 4 \ Na_2SiO_3 \\ 5 \ Fe_2Si_3O_9 \end{cases}$$

about one-sixth of the iron in the first member being replaced by Ca, Mg, and Mn. Riebeckite resembles arfvedsonite, the only other amphibole so rich in iron, in its easy fusibility: splinters of the mineral will melt in a candle-flame.

The other mode of occurrence of riebeckite in the Mynydd Mawr rock is in minute rectangular crystals or short needles with rather high refractive index, pretty strong double-refraction, and nearly straight extinction. They are for the most part colourless, but some give a faint tinge of indigo-blue for vibrations parallel to the long axis. These little crystals are scattered through the ground, and also occur within the porphyritic felspars. They have a fluxional orientation agreeing with the lines of flow marked out by the trains of large crystals. This circumstance negatives the idea that they may be of secondary origin, as supposed by Sauer for the similar microlites in the Socotra rock.

It should be noted that in the Mynydd Mawr rock the riebeckite takes the place, to some extent, of other iron-bearing minerals. Magnetite and augite are both wanting, and biotite occurs only rarely and in minute flakes. The only other locality where I have found riebeckite is at the head of Nant Ffrancon, in another presumed volcanic neck continued into the Bwlch Cywion laccolite. Here, however, it occurs only in the form of minute crystals or microlites. I have not detected the mineral with certainty in any specimens of the rhyolitic lavas: if a more careful search among the higher flows of the Snowdon district should reveal its presence, the point will be of some interest.

Returning to the acid irruptives in general, we notice that in all these rocks the great bulk of the ground-mass consists of felspar and quartz. The former is probably almost exclusively potash-felspar, as appears both from the sections and from the chemical composition of the rocks, taking into account the fact that the bulk of the porphyritic felspars are of plagioclase varieties. The relations of the felspar to the quartz in the ground-mass and of both to the porphyritic elements enable us to recognise among the rocks in question a number of graduated types, or rather a complete series of transitions from a granite-porphyry with granitic affinities to a quartz-porphyry which passes into rhyolite.

In the first type the ground-mass has the characters of a very fine-grained granite. The felspar is fairly well developed and sometimes shews twinning. It often gives squarish sections which ·no doubt belong to orthoclase: occasionally others have the appearance of microcline, and an acid plagioclase is also found in a few cases. The quartz is interstitial to the felspars. Rocks of this kind occur at Trefor ‘and Nant Gwrtheyrn in Yr Eifl, at Mynydd Nevin, etc. [40, 634, 610, 672]. We may call them for distinctness the Nevin type.

A rather different structure is seen in the examples from the neck at the head of Nant Ffrancon, and may be styled the Bwlch-Cywion type. Here the ground-mass is more distinctly ‘hypidio-morphic’ (Rosenbusch), the grains of quartz and felspar interlocking in irregular fashion with no trace of crystal outline [316, 317], and the texture being sometimes rather coarse [685]. Here too the porphyritic crystals are not large or numerous, and except for form there is no great distinction between the earlier and later generations of minerals. In the Mynydd Mawr rock somewhat similar relations obtain, but the texture has sunk to a very fine grain[1] [82, 637]. We seem to have here one line of transition from the coarsely granular types to those with crypto-crystalline ground. In the former the quartz has very rarely separated out in grains at an early stage of the consolidation [620], but as the texture sinks almost to the crypto-crystalline there is sometimes a tendency in the quartz of the ground to collect at particular points,

[1] Professor Bonney ascribes the structure of the Mynydd Mawr rock to devitrification, but here, as in some of the rhyolites, I can find no decisive evidence of an originally glassy state.

forming grains, which are yet not distinct from the ground-mass. This is well displayed in specimens from Carn Fadryn and Carn Fach [22, 624, 625]. It is perhaps not a long step from the Carn Fach type to the crypto-crystalline quartz-porphyries enclosing distinct grains of quartz.

Again, when the separation of the quartz is very imperfect and the ground-mass in general crypto-crystalline, the rocks become identical with the Penmaenbach type of lavas described in an earlier part of this essay [630].

Another line of transition, however, and a very interesting one, depends on the gradual coming in of granophyric structures. Starting from rocks of the Nevin type with an ordinary granular ground-mass, we can trace in the slides from various localities a gradual replacement of the micro-crystalline and micro-granitic structures by micro-pegmatite. A ground-mass composed partly of small felspar crystals moulded by quartz and partly of a pegmatitic intergrowth of the two minerals is found in numerous examples from Trefor, from Craig-dol-Owen (above Aber Cascade), and from some of the hills about Clynog-fawr and between Pistyll and Nevin. In some sections there is very little of the granophyric intergrowth, and that of a very irregular kind [342, 315, 628], but in others it becomes more prevalent and better developed [84, 314]. Sometimes the granophyric structure becomes very minute and the granular part of the ground at the same time excessively fine-grained [629], and in this way there is perhaps a passage again to the crypto-crystalline type of ground-mass. Probably each intrusion contains various gradations of structure, but many of the specimens are spoilt by decomposition and the production of secondary quartz [31, 343, 684, *etc.*].

Next come the typical granophyres. The micro-pegmatitic intergrowth, which in varying degree characterises so many of the acid irruptives of Caernarvonshire, is here developed to such an extent as to form the whole of the ground-mass of the rock. A feature seen in many of the granophyres, as well as in the transitional varieties of rock mentioned above, is the tendency of the micro-pegmatite to grow about the porphyritic felspars with a definite arrangement towards them. On closer examination it may often be verified that the felspathic constituent of the micro-pegmatite is in crystalline continuity with the crystal upon which it grows. This may be observed in granophyres from other countries, for instance from the Lake District [751, etc.], the divisional line of a Carlsbad twin of orthoclase being traceable through the micro-pegmatite surrounding the crystal. In the Lake District grano-phyres, also, as in those of Caernarvonshire, when the micro-pegmatite is arranged with evident relation to plagioclase crystals, the latter are seen to be bordered with a narrow coating of ortho-clase, as described above. In some of the rocks with granular ground-mass granules are often included in the marginal portion of the porphyritic felspars, as if indicating that these continued to grow after the ground-mass had begun to consolidate. In this case, too, the granules of felspar immediately on the edge of a crystal sometimes have the same crystalline orientation (indicated by the extinction-position) as the crystal itself. This and the peculiar disposition of the granophyric growth, to which allusion has been made, appear to signify that no very marked change of conditions divided the earlier from the later phases of consolidation. I am aware that Professor Judd[1] has drawn from similar pheno-mena very different conclusions; but after reading his extremely interesting paper and re-examining my Caernarvonshire specimens, I find nothing in this district to support any theory which ascribes the granophyric structures, the zoning of the felspars, etc., to secondary devitrification. Phenomena indicating a continued growth of the porphyritic felspars during the later stages of the consolidation—a kind of 'original enlargement' of the crystals—have been recorded by various observers in the acid irruptive rocks of many districts[2]. Good specimens of granophyres occur in the Moel Perfedd laccolite, at Y Drosgl and Craig-dol-Owen, at Pen-y-gaer (2½ m. E. of Llanaelhaiarn), and between Pistyll and Nevin. When the structure is on a comparatively large scale, it is rather irregular [611, 670]; when the disposition of the granophyric growth is more regular, it invariably shews the connection with the porphyritic felspars already alluded to [686, 702, 693]. The

[1] Q. J. G. S., vol. xlv., p. 175; 1889.

[2] Fouqué and Michel-Lévy, Min. Microg., pp. 193, 194; 1879. Höpfner, Neu. Jahrb., Beil. Bd. i., p. 164; 1881. Williams, Neu. Jahrb., Beil. Bd. ii., p. 605; 1883. Haworth, Johns Hopkins Univ. Circ., No. 65, p. 70; 1888: Amer. Geol., vol. i., pp. 290—292; 1888. Teall, Brit. Petr. pp. 180, 323, 327; 1888. Cf. Judd, l.c., and Miss Raisin, Q. J. G. S., vol. xlv., pp. 252, 253; 1889.

Moel Perfedd rock is an excellent type. From the proportions of the alkalis in the chemical analysis, it is probable that all the felspar of the micro-pegmatite is orthoclase, and it is noteworthy that with the coming in of the granophyric structure the porphyritic felspars become more exclusively plagioclase.

The normal granophyres (micro-pegmatites) seem to grade into the crypto-crystalline and vitreous quartz-porphyries and rhyolites through rocks with a pseudo-spherulitic ground-mass, but the latter are not of common occurrence. The first step seems to be a diminution in the scale of the granophyric intergrowth, accompanied by an approach to the 'centric' arrangement. A specimen from Foxhall, Llanbedrog, from the Mynydd-tir-y-cwmmwd boss, shews an abundant growth of rather delicate micro-pegmatite bordering the felspars in the usual fashion, but in other parts of the slide are 'ocellar' structures having a nucleus of comparatively coarse granophyric intergrowth surrounded by granophyre on an excessively minute scale with a more or less evident radial arrangement [23]. It would be difficult to draw a line between this structure and the 'pseudo-spherulites' met with in the little intrusion at Penmaen Castle, Pwllheli [458, 647 a]. These growths appear to consist of much-altered felspar with quartz, and have a marked radial arrangement. On the outside of the pseudo-spherulite the distinct fibres may be recognised, but in the inner part it is impossible to say whether the several constituents are really individualised or not. The growth of the structure is not very regular, and is often hemispherical. I have not been able to find *in situ* such true spherulitic rocks as occur in fragments in the agglomerates.

The gradations of structure just indicated are accompanied by increasing rarity of augite, biotite, and magnetite, and the gradual development of porphyritic quartz, which in the crypto-crystalline quartz-porphyries is a constant feature. It occurs in pyramidal crystals, partially rounded as if by magmatic corrosion, and contains yellowish glass-cavities similar to those in the Snowdonian rhyolites. A fluidal structure is also evident in most of the glassy (or devitrified) and crypto-crystalline examples, and there is no reason why they should not be named rhyolites [32, 33, 665]. They resemble specimens from the undoubted lava-flows of the Lleyn district, though, as noted above, the common types of rhyolite in the eastern division of Caernarvonshire are poorer in porphyritic elements.

A description of the granite-porphyries and granophyres would be incomplete without reference to the dark concretionary patches seen in the rocks at several localities. They are usually from one inch to three or four inches in diameter, of ellipsoidal or irregular form, and rather sharply defined. Good examples are seen in the quarries at Trefor in Yr Eifl and at Y Wern near Nevin. Notwithstanding their definite boundaries, these patches are not included fragments[1] but segregations representing an early phase of consolidation, or perhaps brought up from a considerable depth. They closely resemble the dark patches in the Shap Fell granite and others described by Mr J. A. Phillips[2]. Like these, they are finer in texture and more basic in constitution than the surrounding matrix. One slide shews a multitude of little grains of augite [671]. Biotite is very abundant in the form of small but stoutly-built crystals [342, etc.]. Magnetite is very plentiful, sometimes in two distinct generations [16]. Sphene is present at Y Wern [611], and clear apatite needles occur in unusual profusion [342, 611]. There is much less quartz than in the surrounding rock and no granophyric intergrowth. The patches sometimes enclose porphyritic crystals of plagioclase similar to those in the granophyres and granite-porphyries themselves [611].

The gradations of structure and texture which in south-western Caernarvonshire connect rocks of thoroughly intrusive type with equally typical rhyolites undoubtedly point to a relationship subsisting among all the acid rocks of the county. One phase of this relationship has been suggested above, and we shall return to it in the sequel. In the eastern division of the county, we shall see good reasons for believing that some parts of the Y Foel Frâs tract, the eastern end of the Bwlch Cywion intrusion, and the isolated occurrence at Mynydd Mawr represent the sites of old volcanoes from which the Bala lavas of Caernarvonshire were poured out. Other centres of eruption may have existed in the south-western division, but the surviving relics of lava-flows in

---

[1] Specimens from Trefor shew small included fragments as well as the segregatory patches. The former are more sharply bounded than the latter, and have a much more compact appearance: they are doubtless indurated shale.

[2] Q. J. G. S., vol. xxxvi., p. 1; 1880.

that district do not enable us to identify the individual sources
in a satisfactory manner. The defect, however, is more than
counterbalanced by the fact that we can observe there intrusive
rhyolites with apparent transitions into the granophyres and
granite-porphyries.

The mineralogical resemblance between the usual acid intrusive rocks of Caernarvonshire and the rhyolitic lavas is sufficiently close. The most striking points are the occurrence in
both sets of rocks of augite and biotite to the total exclusion of
hornblende (the abnormal riebeckite excepted), and the prevalence
of porphyritic crystals of an acid plagioclase which probably contains some potash. The differences are such as should be expected
between associated volcanic necks and volcanic lavas. The former
are on the whole less 'acid' than the latter, and this is especially
indicated by the greater abundance of iron- and magnesia-bearing
minerals in the intrusive masses[1].

There are other areas of acid rocks in Caernarvonshire which
present many points of interest. Between Caernarvon and Bangor
extends a tract consisting of granites, granite-porphyries, granophyres, quartz-porphyries, rhyolites, volcanic agglomerates, breccias,
and ashes; and a parallel stretch of country between Llanllyfni and
St Anne's Chapel is occupied by rhyolitic quartz-porphyry and acid
fragmental accumulations. All these rocks, however, are of doubtful age, and seem to have no connection with the Bala period of
vulcanicity; and accordingly we shall exclude them from our
subject. The only acid rock then remaining is a granite in the
Lleyn peninsula, which will be briefly noticed.

With the exception of rocks in the neighbourhood of Caernarvon, the relations of which are a matter of debate, and which
are excluded from our subject, the only normal granite in the
county is that which extends for four miles in length and a mile
and a half in breadth on the west of Sarn. This rock builds the
round hill Mynydd Cefn-amwlch, and extends under the lower
ground to the south, as far as Meillionydd-bach: there is also an
outlying patch to the north-east at Pyllau-giach (see map, fig. 4).

[1] Cf. Rosenbusch, p. 338.

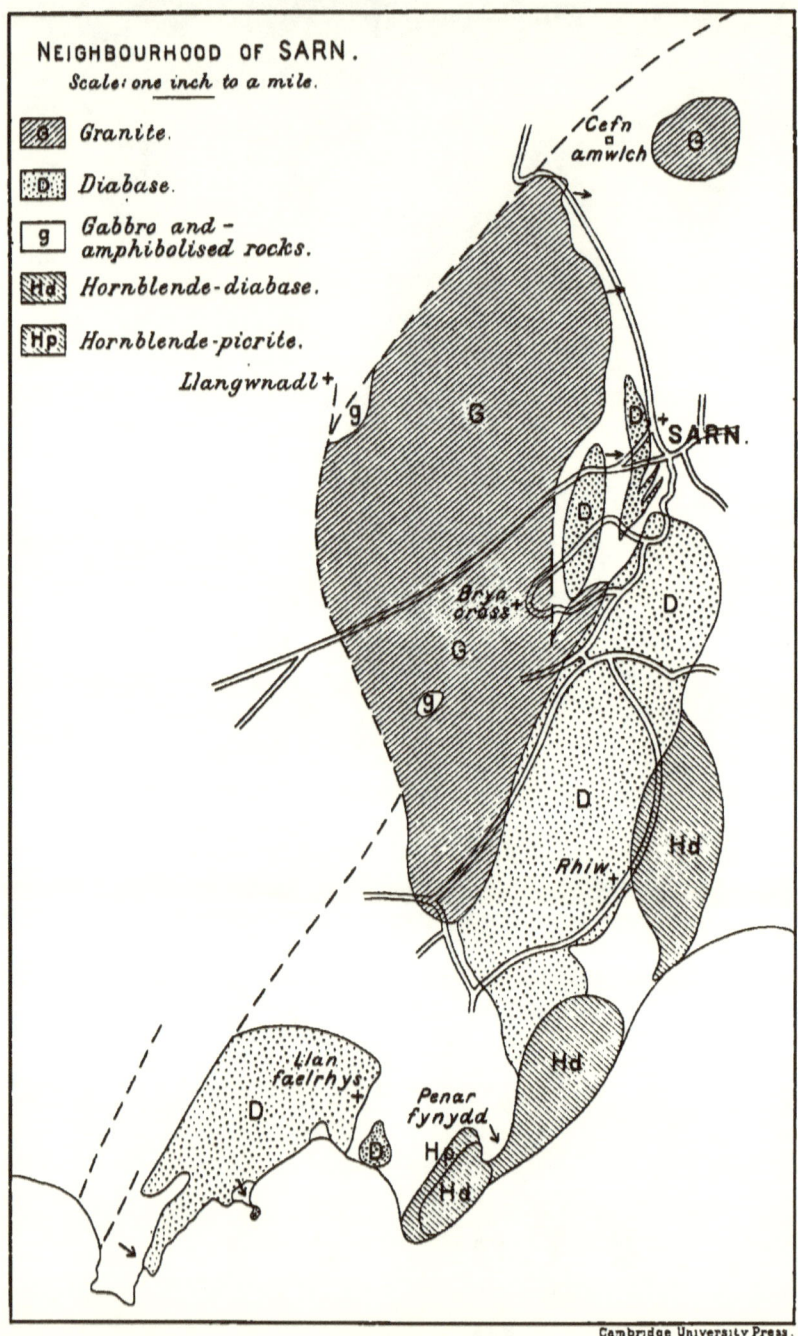

NEIGHBOURHOOD OF SARN.
Scale: one inch to a mile.

G   Granite.

D   Diabase.

g   Gabbro and –
    amphibolised rocks.

Hd Hornblende-diabase.

Hp Hornblende-picrite.

Cefn
amwlch   G

Llangwnadl +

g

g   D +
                          SARN.

G   D

                          D

Brya +
cross +

G

g

                    D

                          Hd

                    Rhiw +

Llan
faelrhys
+   Hd

D   Penar
                    fynydd

        D   Hp
                    Hd

Cambridge University Press.

Fig. 4.

Dr Hicks has claimed this granitic area for his Dimetian system, but I have described elsewhere its field-relations, and shewn from the sections, both at Cefn-amwlch and at Meillionydd, that the mass is intruded through the adjacent strata, which are of either Upper Arenig or Lower Bala age. Professor Blake, also, has with-drawn his opinion which referred the granitic rocks of this district to his Monian system.

Sir A. Ramsay (p. 220 and *Catalogue*, p. 41) names the rock 'syenite,' and describes it as containing hornblende; but this mineral is decidedly exceptional, and the ordinary type of rock is a biotite-granite or 'granitite' of some authors. To the eye it presents an appearance similar to that of the St David's rock and some other ancient granites, the peculiarity arising from the abundance of quartz, the rather small proportion of the ferro-magnesian constituent, and the usually advanced decomposition of the latter.

The continuity of the granitic mass between its northern and southern exposures is rather inferred than demonstrated, but the practical identity of the whole cannot be doubted. Although the Meillionydd House granite exhibits a certain parallel and even gneissose structure, apparently due to fluxional movement near the margin of the mass, the rock seen in the little hill immediately to the east is precisely of the same type as that quarried near Cefn-amwlch and at Pen-y-gopa. The mapping of the Geological Survey is at fault here, much of the granite in the main mass and the whole of the Pyllau-giach occurrence being coloured as 'green-stone,' and a tongue of granite at Mûriau near Bryn-crocs being omitted, while on the other hand the diabase hill Clip-y-cilfinhir is mapped as granite. (*Cf.* Hicks, Tawney, and Blake.)

Junction sections shew not only a pronounced induration of the shales, but a certain amount of modification in the granite itself, which developes segregatory nests of dark mica, as is seen at Meillionydd and between Ty-rutten and Penllech. In the former locality, too, there is a kind of gneissic structure, evidently a fluxional phenomenon.

The microscope shews the usual constituents, and as a rule the normal structure, of a biotite-granite. Iron-ores are not usually met with in any quantity, but a little titaniferous magnetite is sometimes to be noticed. Apatite is rarely seen. In the Meillionydd rock Mr Teall (p. 320) recognised crystals of zircon, which, when they occur in the mica, are surrounded by the characteristic aureole of strong pleochroism.

Biotite occurs in flakes, sometimes bent or yielding along *gleitflächen.* The absorption is as usual $\gamma = \beta >> \alpha$;

　　α, rather pale yellow-brown;

　　$\beta = \gamma$, very intense brown, rather greenish, and sometimes almost opaque.

Some altered portions are greener, and give

　　α, pale brownish-green or greenish-brown.

A little decomposing green hornblende accompanies the mica in a few cases.

Two felspars are always present, as crystals of more or less rectangular outline. The first is an acid plagioclase giving low extinction-angles, very finely lamellated and sometimes shewing a second (pericline) twinning. The other felspar appears to be orthoclase, and is usually untwinned. Of these the former, probably oligoclase, is always the earlier formed, and in general the more abundant. Both species are deeply affected by the so-called kaolinisation; which sometimes begins in the interior, but frequently proceeds in a capricious fashion leaving parts of the crystal quite clear.

The quartz forms, as a rule, crystalline grains moulding the felspars, but also not infrequently included by them. In other slides it is partly in the form of an irregular mosaic. Connected perhaps with the rather unusual abundance of the mineral is its tendency to crystallise in part before the orthoclase, sometimes even in good dihexahedra. The quartz of the rock is usually crowded with minute cavities containing spontaneously moving bubbles [632, 635, 704]. Often these cavities are ranged in roughly parallel planes, which pass through several contiguous grains of a quartz-mosaic, and in this case may plausibly be referred to the action of strain on the rock-mass after its consolidation[1].

The rock quarried near Meillionydd House shews a rather unusual structure, almost the whole of the biotite being of later consolidation than the felspars and quartz, which it moulds and

---

[1] *Cf.* Judd, *Min. Mag.*, vol. vii., pp. 81, *etc.*; pl. iii., fig. 1; 1887.

even encloses. This gives a curious ophitic appearance to some sections as viewed under the microscope [814, 635].

Specimens from all parts of the mass often shew signs of crushing. The 'gliding-planes' in the mica, the occasional bending of the felspar crystals, and the rows of fluid-cavities in the quartz seem to point to mechanical stress. Again it is not uncommon to see a section traversed by veins of quartz and felspar mosaic with granules of epidote, and in some places crystals of the constituent minerals of the rock are surrounded by a fine-grained aggregate of the same, in a manner recalling the *Mörtelstructur* described by Törnebohm [144, 704, 705].

There is not sufficient evidence to fix with certainty the age of the granite, or to suggest possible relations between it and other igneous rocks in the district. The elongated form of the main mass and of the little offshoot at Mûriau seem to indicate that the molten rocks were injected with a certain rough parallelism to the bedding-planes of the strata, and other appearances tend to shew that the intrusion took place prior to the cleavage of the neighbouring shales. The balance of evidence is therefore in favour of assigning the granite to the Bala age, and this view is confirmed to some extent by the mineralogical resemblance of the granite to the various acid rocks, intruded and extruded, of the Bala volcanic age in Caernarvonshire. The boundary of the mass on its north-western and the greater part of its western side appears to be faulted.   ＼

V. Intermediate Rocks.

Rocks of 'intermediate' composition occur in Caernarvonshire only in a few scattered exposures, and, with the exception of the one at Penmaenmawr, have scarcely received notice. They possess, however, a certain interest, both in themselves and in relation to the other igneous rocks in the county, and such examples as I have been able to find in various parts of the county will be briefly described. The well-known quartz-bearing rock of Penmaenmawr will be the first considered.

The rocks forming the heights of Penmaenmawr and the smaller hill half a mile to the south, above Tai-rhedyn, present an uncommon and interesting petrological type or series of types. They were described as 'felspathic porphyry' by Sir A. Ramsay (p. 137), who commented on their probable relation to the adjacent strata. The black slates are much altered at the contact; "and though the form of the ground seems to indicate that the rock of Penmaenmawr rests upon the slate with a northern dip, yet the massive and crystalline character of the rocks,...and their boss-like forms, seem to shew that they are intrusive masses, and not interbedded felspathic lava-flows." It is very likely, indeed, that better exposures of the junctions would prove these two masses to be large laccolites injected among the Arenig or Lower Bala strata. It is even possible that the two are remnants of one large mass bent over the intervening anticlinal, where the slates have a well-marked cleavage striking E. 15° N.—W. 15° S. This, however, is only a conjecture, and the Survey map shews the southern intrusion impinging on the lowest lava-flows. In the maps and memoir of the Geological Survey the rock of Penmaenbach is included with those mentioned above, but as its field-relations appear to be quite different, and the rock itself is a rhyolite, it has been separately considered in its place among the acid lavas.

Penmaenmawr rises to a height of 1550 feet, and the uppermost 500 feet have been extensively cut into by quarrying operations. The rock was first described as a quartziferous diorite by Mr J. A. Phillips, who made analyses of the fresh rock (I.) and also of altered specimens (II., III., IV.).

The first analysis is sufficient to shew that the rock belongs to the 'intermediate' family. The alteration experienced by the other specimens is due to ordinary weathering agents, and consists mainly in the removal of part of the lime and magnesia, and an increase in the percentage of water: the quantities of the other substances present are only *relatively* increased. These changes are accompanied by a decrease of density.

	I.	II.	III.	IV.	V.
SiO_2	58·45	60·31	62·24	61·75	65·1
Al_2O_3	17·08	18·99	18·25	18·88	12·9
Fe_2O_3	0·76	1·07	1·05	0·52	2·0
FeO	4·61	4·31	3·08	3·52	4·7
MnO	trace	trace	trace	trace	trace
CaO	7·60	5·81	4·69	3·54	4·7
MgO	5·15	0·83	2·27	1·90	2·8
K_2O	1·02	1·67	1·49	1·24	3·9
Na_2O	4·25	4·55	3·79	3·67	2·8
H_2O {hygro.	0·12	0·40	0·05	traces}	1·9
{comb.	0·95	1·82	2·64	4·46}	
P_2O_5	trace	trace	trace	trace	—
FeS_2	—	—	0·39	0·09	—
	99·99	99·76	99·94	99·57	100·7
Sp. gr.	2·94	2·79	2·75	2·79	2·72

I., II., III., IV. J. A. Phillips; *Q. J. G. S.*, vol. xxxiii., p. 424; 1877.

V. T. H. Waller; *Midl. Naturalist*, 1885, p. 5.

Phillips' paper was published in 1877, and in the same year Rosenbusch[1] referred to 'the traps of Conway and Penmaenmawr'

[1] *Mikr. Phys. d. mass. Gest.* p. 352; 1877: 2nd ed., p. 204; 1887.

as typical examples of the enstatite-bearing diabases. The mineral formerly taken for hornblende was found to be one of the enstatite family and to have the optical properties of bronzite. It is accompanied in the rock by augite. This was confirmed by Mr Teall[1], who also noted the occurrence of quartz in the form of micropegmatite in the basis of the rock.

Mr Waller subsequently made a closer examination of the Penmaenmawr rock, and paid special attention to certain light-coloured veins which frequently occur in it. His analysis of the material of the veins is cited (in column V.) for comparison with the normal rock (I.) of the same quarry. It will be seen that the vein contains higher percentages of silica and potash, with considerably less lime and magnesia, than the typical rock. These facts accord with the idea that the veins represent the latest product of consolidation of the rock-magma. More recently Mr Teall (pp. 272—276) has given a description of the Penmaenmawr rock under the title 'enstatite-diorite,' and has discussed the significance of the veins.

As seen in the field, the typical rock is of a bluish or greenish-grey colour, and of fine to medium grain, with a speckled appearance due to the mingling of felspar and pyroxene; but there is also a very compact variety with visibly porphyritic felspars. This latter, according to Mr Waller, forms the marginal portion of the Penmaenmawr intrusion in the western quarries; it is also found on the southern side of the hill, and constitutes the upper part of the Tai-rhedyn mass. The veins are of a light-grey or white colour and partly coarse-grained, with conspicuous slender crystals of black augite. They are often lenticular in shape, with a parallel arrangement, and when numerous give a singular streaked aspect to the rock.

Having regard to both the mineralogical constitution and the structure of the chief type of rock, it is perhaps best described as a bronzite-bearing quartz-dolerite. The compact type, with frequent fluxional arrangement of the included felspars, may be named quartz-andesite. The majority of the specimens microscopically examined belong to the doleritic type, although the structure sometimes passes into that of a quartz-diabase by merging of the different generations of felspar.

[1] Q. J. G. S., vol. XL., p. 656; 1884.

The minerals present are apatite, magnetite and more rarely ilmenite, bronzite, augite, biotite, felspars, and quartz. Slender needles of apatite and small magnetite crystals are always present, and represent the first results of the consolidation. Occasionally are seen parallel plates of ilmenite, or perhaps an intergrowth of that mineral with magnetite [318, 574].

Bronzite, or a mineral pseudomorphing it, is almost always present in the slides. It builds idiomorphic crystals, often rather rounded at the edges, the forms seen in a cross-section being the two pairs of pinacoids well developed and truncated by the prism-faces. The prismatic cleavage is well marked. The colour is pale brown with a slightly greenish tinge, and in longitudinal sections a slight dichroism can be verified. The bronzite, however, is often converted into a pale-green, rather fibrous mineral resembling serpentine and identified by Rosenbusch as bastite. This is pleochroic and doubly refracting, giving extinctions parallel and perpendicular to the length of the pseudomorph.

Augite always accompanies the rhombic pyroxene, and in about equal quantity. It closely resembles the other, but is distinguished by the oblique extinctions in longitudinal sections, the absence of pleochroism, and a delicate striation parallel to the basal plane, as remarked by Mr Teall. The augite does not usually present such good crystal outlines as the bronzite, but occurs for the most part in rather rounded grains [320, etc.], and more rarely in small sub-ophitic plates. Twinning is not uncommon [323], and an intergrowth of the two pyroxenes is sometimes observable [574].

Flakes of biotite are often seen in the slides, and sometimes in abundance [321]. The mineral has the usual brown colour and strong dichroism, though in some examples it is partially bleached [320, 323].

In most specimens three generations of felspars are readily distinguished [320, etc.]. The first consists of long irregularly rectangular crystals some 0·05 inch in length, with Carlsbad and albite-twinning, and with extinction-angles which seem to point to labradorite. These often include the rounded grains of augite, although some portion of the latter mineral seems to be later than these felspars. Next we have numerous smaller rectangular felspars with albite-lamellation and sometimes Carlsbad or pericline-twinning in addition. These seem to belong to the andesine-oligoclase series. Lastly, there are allotriomorphic felspars, clear and with distinct cleavage-traces, of which a part at least must be referred to orthoclase. These last, with the quartz, make up the ground-mass of the rock. The two minerals are intimately associated, with every gradation of the granophyric structure, many of the specimens affording most beautiful examples of micro-pegmatite[1] [318, 323, 574].

Some slides exhibit a rather different ground-mass, consisting of crowds of little lath-shaped felspars, twinned once or more, with small granules of augite and an interstitial paste of clear quartz [321].

The andesitic type of rock is clearly linked to the doleritic by its general characteristics. The slides shew, in some specimens, porphyritic felspars like those described above, with crystals of feebly dichroic, brownish bronzite and a few of augite [684]. Sometimes augite alone is present, in grains and imperfect crystals. The ground-mass consists partly of felspar microlites, but there is also an isotropic base, and the quartz is scarcely individualised [669].

The rocks next to be described are pyroxene-andesites without quartz. Such rocks occur, though only slightly developed, in the Lleyn peninsula, but only one example, that of Carn Boduan, has previously been recorded. In petrological character these rocks are thoroughly volcanic, but their geological relations are rather obscure. In one or two places they are clearly interbedded lava-flows occurring among Bala strata, and possibly careful mapping might shew that they have a greater extent laterally than I am now able to assert, and that they are lavas occupying a definite horizon below the chief rhyolites of the district. At least it is worthy of note that the four outcrops mentioned below occur very nearly at one stratigraphical position, being situated on a belt encircling the little synclinal trough of central Lleyn.

The augite-andesite of Penmaen Castle, a mile west of Pwllheli (see map fig. 3), is a true lava-flow. It is associated with and underlain by agglomerates, consisting largely of fragments of the same rock but enclosing also pieces of indurated black shale. The rocks dip N. by W. at 60°, passing under the rhyolites, from

---

[1] See Teall, plate xxxv. fig. 2.

which they are separated by a small thickness of shale, while immediately to the south is an intrusion of spherulitic quartz-porphyry. All these rocks are marked by one red tint on the Survey map.

A similar augite-andesite is seen in Y Foel Fawr, a mile and a half west of Llanbedrog, but whether breaking through the adjacent rhyolites or interbedded with them and itself broken by intrusive rhyolites I cannot venture to say. The rock is not distinguished by the Survey. (See map fig. 2.)

Farther to the north-west, at Bodlas, about a mile and a half above Nanhoran, is another exposure, which seems to be nearly on the line of the last one. The rock is much weathered, and has been mapped as intrusive greenstone.

Finally we have the conical wooded hill Carn Boduan, a mile south of Nevin, which is clearly distinct from the neighbouring granophyres, granite-porphyries, etc., although it has been included with them, both by the Survey (as 'intrusive felspathic rocks,' Ramsay, p. 219), and by Dr Hicks (as metamorphic 'Arvonian'). The latter author asserts that the rocks are "in the main lava-flows, breccias, or hälleflintas." Professor Bonney, in an appendix to Dr Hicks' paper, describes a specimen as either a quartz-felsite or a quartz-porphyrite. Mr Tawney[1] collected specimens from various parts of the hill, and stated that the whole is of porphyrite. As confirmatory of its igneous and possibly intrusive nature, he mentions an included fragment of much indurated shale with crystals of magnetite, mica, and garnet, which he found at the summit of the hill [622]. Professor Blake apparently considers the rock a lava, and states that it is "succeeded by a slate containing its pebbles, obtained by contemporaneous erosion."

The general appearance of the mass seems to be rather that of a boss, or perhaps a neck, intruded through both the slates and the acid eruptives. I find no breccias, nor any sign of the bedded appearance spoken of by Dr Hicks, unless a well-defined platy jointing can be so designated. Whatever be the relations of the rock, specimens from various parts of the hill are undeniably andesites, and of an interesting type. It will be convenient to separate it from the other andesites, which will be described first.

[1] *G. M.*, 1882, pp. 549—552.

The andesites of Lleyn, then, excluding the Carn Boduan rock, are augite-andesites offering few peculiarities. In hand-specimens they have a compact appearance, and the colour of the freshest examples is dark grey to olive-green. The rocks from Penmaen Castle and Bodlas, including the angular fragments in the breccia at the former locality, are vesicular. The numerous vesicles, about a quarter of an inch in diameter, are usually spherical but sometimes of irregular forms, and are filled with a black substance, apparently chlorophæite.

Thin sections shew a fine ground-mass enclosing porphyritic felspar, augite, and occasionally grains of magnetite. The augite is of prior consolidation to the felspar, which often moulds it. It forms colourless grains with a somewhat rounded outline. In the Y Foel Fawr rock [681] most of the augite is pseudomorphed by green and greenish-brown hornblende. The porphyritic felspars are about 0·1 inch in length and of elongated rectangular form. They shew the usual lamellation, rarely crossed by pericline-twinning, and from their extinction-angles may be referred to andesine.

In a specimen from directly under Penmaen Castle [605 a] the distinction between the two generations of felspars is not very clearly marked. The crystals range from 0·04 to 0·002 inch in length. The larger ones are twinned as described above; the smaller are twinned once only. In the other specimens, however, the later felspars consist of microlites which make up a large portion of the ground-mass. The interstices are filled by an isotropic base, which was probably vitreous. There is always a beautiful fluxion-structure marked by a parallel arrangement of the small felspars.

The chloritoid substance which fills the vesicles already noticed in some of these andesites [579, 605 a], is, in transmitted light, of a pale yellowish-green tint, but often almost colourless at the border. Much of it is sensibly pleochroic; that in the central part of the cavities is often very weakly doubly-refracting, and has a radial-fibrous arrangement. Veins of a similar substance, with confused structure, traverse the rock in some places. The optical properties of the minerals belonging to the chloritoid or saponite family are unfortunately but little known. So far as we can judge, these vesicles appear to be either filled with chloro-phæite, or lined with that mineral and filled in with delessite.

Coming now to Carn Boduan, we find that the rock, when freshest, is black and compact, shewing to the eye small imbedded felspars with twin lamellation. Most of the exposed rock, however, is grey, and it even becomes yellow and brown. Quartz is liberated during the weathering, as well as calcite [67 a, 68], and sometimes veins of chalcedony are formed, as in the quarry near the turnpike, where all the varieties are well shewn. The quartz seen in slides cut from the weathered rock is absent in fresh specimens, and is certainly of secondary origin [67 a, 69]. The mineral referred to hornblende by Mr Tawney is an altered rhombic pyroxene, and as it is accompanied, though in less quantity, by augite, we may name the Carn Boduan rock a pyroxene-andesite.

The specific gravity of a fresh specimen from the quarry is 2·646, and an analysis kindly performed by Mr Acton is given below (I.). The iron is all estimated as ferric oxide. For comparison I give the analyses of two of the Lake District andesites

	I.	II.	III.
SiO_2	61·8	60·718	59·511
Al_2O_3	16·5	14·894	17·460
Fe_2O_3	6·7	1·405	1·271
FeO	not est.	6·426	4·926
CaO	4·5	6·048	5·376
MgO	1·2	1·909	1·801
K_2O	1·4	2·354	3·705
Na_2O	7·2	2·843	3·093
Loss on ignition	0·6	0·964	0·483
FeS_2	—	0·395	0·604
P_2O_5	—	0·281	0·115
SO_3	—	0·103	0·086
CO_2	—	1·660	1·569
	99·9	100·000	100·000

I.   Carn Boduan.   Anal. E. Hamilton Acton.

II.   Brown Knotts.   Anal. J. Hughes; *Q. J. G. S.*, vol. xxxi., p. 408; 1875.

III.   Iron Crag.   do.   do.

(Keswick series), which have close resemblance to° that now described (II. and III.). The chief difference observable between the columns is in the alkalies, the Carn Boduan rock having a large preponderance of soda over potash, while in the Cumberland lavas the quantities of the two alkalies are almost equal. These latter rocks, too, are manifestly rather decomposed, and the figures are evidently calculated up to a total of 100. The Carn Boduan rock does not differ much, except as regards the alkalies, from the Penmaenmawr quartz-dolerite, of which analyses have been cited above.

The structure and constitution of the rock are beautifully exhibited in some of the slides [642, 643]. The constituents, in the order of their separation, are apatite, magnetite, hypersthene, the earlier felspars, augite; then the later magnetite, felspar, and augite, with some residual base.

Apatite occurs sparingly in the usual slender prisms; magnetite in crystals in the form of rather rounded octahedra.

The mineral which is here considered to represent a rhombic pyroxene is for the most part completely pseudomorphed by a green serpentinous product, preserving the form of the rectangular crystals, and having a fibrous structure parallel to their length. This substance gives straight extinction, and is sensibly pleochroic or dichroic, vibrations parallel and perpendicular to the long axis (c) giving respectively a light bluish-green colour and a very pale yellowish-green [642]. This mineral is indeed closely similar to the 'bastite' of Rosenbusch in the Penmaenmawr rock described above, and there can be no doubt that the crystals which it pseudomorphs belong to bronzite or hypersthene. A precisely similar replacement is seen in the rhombic pyroxene of the Shropshire andesites[1] and in some of those of the volcanic series in Cumberland [746, etc.].

The porphyritic felspars are plentiful, with a tendency to collect in clusters, and occasionally a slight parallel arrangement. The crystals present rather elongated sections, for the most part finely lamellated, and sometimes exhibiting cross-lamellæ. The extinction-angles point to a variety near andesine. In some cases, however, the angles are lower, and this variety probably ap-

[1] I am indebted to my friend Mr W. W. Watts, who has shewn me his collection of these rocks.

proaches oligoclase. These latter crystals are rather less elongated and of less regular outline than the former. If they are monoclinic, which is unlikely, their zonary banding, often very pronounced, must indicate varying admixtures of potash and soda ، felspars. In any case, the fact of the later felspars being more acid, apparently, than the earlier is in accordance with the usual rule.

Augite is much less plentiful than the presumed hypersthene ; and it is to be noticed as a characteristic feature that it is of posterior formation to the felspars, or at least to the dominant felspar, while the rhombic pyroxene is invariably earlier [643]. The augite is a very pale yellowish-brown variety in rounded composite grains.

The minerals named above are imbedded in a ground-mass in which are recognised, in order of consolidation, small magnetite granules and crystals, little microlites of felspar, and probably augite granules, with a certain amount of isotropic residue. The microlites frequently shew a beautiful fluidal disposition in lines winding round the porphyritic elements. These felspars of the ground-mass appear to be of a rather acid nature, and probably they, with the isotropic base, contain most of the potash present in the rock. Some slides [631] shew a rather different type of structure, the greater part of the rock consisting of closely packed felspar crystals with a partial parallel disposition.

To the andesitic rocks described above may be added certain others whose affinities with the foregoing are unmistakable, although the specimens are too much decomposed for their minute description to be profitable. These form parts of the large igneous patch, three and a half miles in diameter, shewn on the maps in the neighbourhood of Y Foel Frâs. They are indeed a portion of the ' hornblendic greenstone ' of Sir A. Ramsay (p. 139), which he considered to pass gradually into the 'felspathic porphyry ' constituting the major part of the area.

Specimens from Bera Mawr [312, 313] appear to have been bronzite-bearing andesite and dolerite. The baštite pseudomorphs are precisely similar to those in the Penmaenmawr and Carn Boduan rocks, and the other characters present a close parallel. Here, as in the other cases, the bronzite crystals were idiomorphic and of earlier formation than the accompanying augite. The slides contain quartz, but this mineral appears to be a secondary product. Its presence or absence as an original constituent is perhaps connected rather with the conditions of consolidation than with the exact composition of the magma. It may be noticed that at Penmaenmawr the andesitic type has much less quartz than the doleritic, if any, and further that the quartzless andesite of Carn Boduan has a composition closely resembling that of the Penmaenmawr dolerite, being in fact slightly more acid.

The border of this igneous area to the west and south of the Bera Mawr exposures, also mapped as 'greenstone,' I have not visited, and of the other patch, to the north of Y Foel Frâs, I have examined only the rocks bordering Llyn-yr-afon. These are not all of one type. Those immediately south-east, south, and south-west of the lake are ophitic diabases not in any way different from the others to be described below. Above the western and north-western shores, however, the crags consist of agglomerates and ashes of andesitic appearance [857, 858]. The presence of these fragmental rocks in this place is not easy to explain except on the theory, for which we shall find further warrant, that an earlier volcano with some of its ejectamenta has been swallowed up by a later and larger one, of which this large area of igneous rocks marks the site.

It is not a little interesting to observe the close similarity of the bronzite- or hypersthene-bearing andesites of Bala age in Caernarvonshire, Shropshire, and the Lake District to the Old Red Sandstone andesites of the Cheviots, the Permian rocks of the Nahe district, and the recent lavas of Santorin.

We shall conclude this part of our subject by noticing certain plutonic rocks which are perhaps best grouped under the 'intermediate' division. They are the only igneous rocks in our north-western division of Caernarvonshire which probably belong to the Bala age, and build three intrusive masses breaking through Arenig strata near Llanfaglen to the south-west of Caernarvon. Their intrusive nature is proved by the contact-alteration of the adjacent rocks: the shales at Cefn-coed are baked to a red colour and partially calcined. The outcrops appear on the map as irregular ovals having their long axes in the direction of strike of the strata; and this may be taken to indicate a general approach to the characters of laccolites, although at the observed junctions the intrusions break more or less freely across the strata. The largest mass is about a mile and a quarter in length: a smaller one lies to the west, at Llanfaglen[1] old Church, and another to the north at Glan-y-mor. These rocks, although noted by Sir A. Ramsay (p. 205) and marked on the Survey map, have not yet been described. The three masses present precisely similar varieties, and may therefore be treated together.

The variation of aspect in these rocks as seen in the field is very striking. Some specimens, especially from the eastern portion of the Cefn-coed mass, are almost black, and to the eye shew little beyond hornblende and a few crystals of pyrites. The more usual type, however, has plenty of felspar, though the proportion of this mineral varies much in short distances. The felspar is either white or red, and both kinds may be observed in different parts of a hand-specimen. Flakes of mica are often visible: they seem to be closely connected with the hornblende and to lie parallel to its cleavage-planes. On the whole, the rock is perhaps best classed with the syenites, although it contains some quartz and a considerable quantity of augite. We may speak of it, then, as a quartz-augite-syenite: it is mapped as 'greenstone' by the Geological Survey.

The microscope discloses apatite, magnetite, ilmenite, pyrites, two species of alkali-felspars, augite, hornblende, biotite, and quartz, besides leucoxene, kaolin, a chloritoid, epidote, and calcite.

Apatite, in clear needle-like prisms, is a plentiful though local constituent: as usual it is the first-formed element of the rock [644, 647]. Skeletons of ilmenite [644] and magnetite [646] abound in some slides, while others have no iron-ores except obviously secondary magnetite-dust. Pyrites is not uncommon [644], and is often seen with the naked eye.

The felspars are often in well-bounded crystals, and for the most part clearly of earlier consolidation than the hornblende, which commonly moulds and even encloses them [645, 647]. They belong mostly to an acid plagioclase, but in part to orthoclase, and the crystals are invariably much decomposed.

An augite, colourless in thin sections, is generally present in some abundance, forming crystalline grains, or sometimes im-

---

[1] Llanfaglan on the Survey map; but the name is a contraction of Llanfagdalen or St Magdalen's.

perfect crystals, of earlier consolidation than the felspars [646]. This augite is often bordered by brown hornblende, which may be an original growth in crystallographic relation with the earlier mineral [644]. At other times the pyroxene seems to pass into green fibrous uralite.

Besides the above bordering or 'complementary' and the uralitic hornblende, there are plenty of crystals of the brown 'basaltic' kind, often shewing the prism-form truncated by the clinopinacoid [644, 646]. It has the usual absorption-formula $\gamma > \beta >> \alpha$. Sometimes its colour is greenish-brown or even green, but this is evidently the beginning of a secondary change. The hornblende, as well as the augite, is often converted into a feebly-polarising chloritoid. In other places, probably in conjunction with the felspar, it has given rise to the formation of bundles of epidote crystals [647]. Still another mode of alteration is that which has resulted in the formation of a dirty-brown mica.

Quartz is present in all the slides. From the manner in which it replaces decaying felspars and hornblende, it cannot be doubted that much of this mineral is secondary [644]; but there is always primary quartz in addition. In a few places hexagonal sections are seen [646], but the quartz is usually interstitial relatively to the other constituents. In one slide from Cefn-coed [645] there is a kind of ground-mass composed of a micro-pegmatite of quartz and orthoclase, but this is rather an unusual feature in the rocks.

The general structure of the rock, the behaviour of the iron-ores, the ophitic occurrence of the hornblende, the varying relations of that mineral to the augite, with the presence of the latter constituent, all conflict with the usual characters of a quartz-syenite, and shew some curious points of resemblance to the hornblende-diabase laccolites of Caernarvonshire and Anglesey to be mentioned below.

The date of intrusion of the Llanfaglen syenites can only be conjectured. We know only that they are post-Arenig[1]. I shall, however, refer again to the question of the age of these rocks in connection with the probable sequence of events during the great age of igneous activity in Caernarvonshire.

---

[1] Marr, *Q. J. G. S.*, vol. xxxii., p. 134; 1876: also Ramsay, p. 204. The shales were formerly described as Lingula Flags (*Catalogue*, p. 40), but no strata of that age have yet been recognised in this part of Caernarvonshire.

THE rocks which next demand notice are the diabases. On looking at a geological map of Caernarvonshire, or indeed of North Wales, one cannot fail to be struck by the remarkable coincidence between the distribution of the 'felstones' and that of the 'green-stones.' In the eastern division of Caernarvonshire especially—and for the moment the Lleyn district may be left out of con-sideration—the basic rocks hardly ever occur except in close association with the rhyolitic lavas, and particularly with the later flows. A closer examination, however, reveals the fact pointed out by Sedgwick, and so admirably proved by the work of the Geological Survey, that while the acid rocks form large con-tinuous sheets which have certainly been poured out in the form of true superficial lava-flows, the diabases, on the other hand, have in-variably been intruded among preëxisting lavas, ashes, and sedi-mentary strata, without ever reaching the surface. No evidence of any true lava-flow of basic composition has been discovered in Caernarvonshire[1]. The diabases occur, in fact, in the form of sheets or 'sills' of varying thickness and extent, often following very closely the bedding of the rocks among which they have been injected, but in places breaking across to a new horizon. Occasion-ally the sheets swell into larger masses and assume rather irregular forms, becoming at the same time more clearly transgressive; again, they may take a general lenticular shape while still occupy-ing a definite horizon between the strata; but as a rule the tabular form is well maintained. The contact-alteration of the beds above and below can often be clearly verified.

[1] We reserve the possible exception of the basaltic rocks at Porth-dinlleyn.

It is easy to fix limits to the date of intrusion of these rocks, and the evidence is conclusive that all the diabases associated with the Bala strata of eastern Caernarvonshire were injected during the Bala age itself. It is very noticeable that although these scattered inclusions approach close to the border of the Silurian[1] formations, not one of them penetrates Silurian rocks. Besides this, it is manifest that the intrusions preceded in general the crust-movements which so profoundly affected this district about the close of the Bala age[2]. In many of the sedimentary rocks the bedding-lamination became at that epoch almost completely obliterated by a newly-impressed cleavage-structure; but the diabase sills are always guided by the bedding, never by the cleavage. These sheets have everywhere partaken of the contortion of the strata among which they lie, and it is even possible in the case of the larger intrusions to trace the manner in which the folding has been modified by the presence of these stubborn masses of rock among less resisting materials.

It may be asserted with some confidence that the latest of the basic sheets were injected almost at the time when the plication of the strata was in progress, so that their form and distribution are partly determined by the folds. Many of the sheets situated at the higher horizons in the eastern part of the county are, in fact, rather to be regarded as somewhat attenuated laccolites lying in the troughs of synclinal folds or sometimes occupying the summit of an anticlinal dome. Numerous instances are seen between Capel Curig and Llanrwst and in the country farther north, about Lakes Cawlyd and Eigiau : their relations are clearly indicated on the Survey maps. The injection of igneous rocks into the cavities which tend to form during the folding of strata is probably a not uncommon phenomenon : it has been described in the Corndon district, Shropshire[3], and the Leadville district, Colorado[4], and the laccolites of Wexford and Wicklow[5] are perhaps an instance of the same action.

---

[1] This name is restricted to the strata from the Mayhill to the Ludlow inclusive.

[2] Cf. Sharpe, Q. J. G. S., vol. II., p. 811; 1846.

[3] Watts, Rep. Brit. Assoc. for 1886, p. 670.

[4] Bulkley, Trans. Amer. Inst. Min. Eng., vol. XIII., p. 384; 1885: Neu Jahrb. f. Min., &c. 1888, vol. I., p. 411.

[5] Kinahan, G. M., 1881, p. 134.

A significant circumstance in connection with the foregoing remarks is the almost complete absence of dykes from the district under consideration. Excluding, as before, the Lleyn peninsula, hardly any instances can be cited of dykes associated with the Bala lavas and the diabase sheets. Dykes of basic material do indeed occur in northern and central Caernarvonshire, but we shall see that these were injected at a much later epoch, and their course determined by joints formed during post-Carboniferous crust-movements. When the Bala strata were injected with the diabasic magma, no such joints existed in the sedimentary rocks, and the igneous material accordingly forced its way along the surfaces of weakness offered by bedding-planes. The few exceptions strongly confirm this view, for they are almost exclusively confined to the diabases intruded into the highest strata in the Llanrwst district, which intrusions we suppose to have proceeded concurrently with the beginning of post-Bala folding in this area.

In Lleyn, too, the diabase sometimes forms dykes cutting nearly perpendicularly through the strata; as at Llangian and Llanengan near Abersoch, and about a quarter of a mile west of Sarn. A large number of dykes are marked on the map in the green schists at the extremity of the peninsula; but it is probable that most of these belong to the post-Carboniferous group of intrusions, to be discussed later, and we shall not include them in this place.

We also exclude from present consideration certain laccolitic intrusions of hornblende-diabase, which occur in the neighbourhood of Penarfynydd and Rhiw and near Clynog-fawr.

Although free from olivine, the Caernarvonshire diabases are decidedly basic in constitution. In a specimen from Pant Evan near Tremadoc Mr Acton finds 47·4 *per cent.* of silica, and the percentages of the alkalies (potash 1·5 and soda 3·9) are quite in accord with the normal basic character of the rock. A little titanic acid is present, but no chromium or manganese. Phosphoric acid (0·2) is present as apatite, while water (2·8) and carbonic acid (1·6) indicate some amount of secondary change. The specific gravity of this rock is 2·900, which may be taken as an average figure for the diabases of the county.

Petrologically the rocks under discussion present few features of novelty. They have attracted little attention, and almost the only published notices of them are those contained in Mr Tawney's

Woodwardian Laboratory Notes, and Mr Teall's *British Petrography*. A few of the Lleyn rocks have been described by
Mr Elsden[1] and by the writer[2].

To the eye the rocks appear as ordinary ophitic diabases
stained with the greenish hue which comes from chloritoid and
other decomposition-products. Sometimes they are spotted with
little black patches, seemingly delessite (well seen between
Aberdaron and Pen-y-cil): and in exposed places the destruction
of these spots or of augite grains gives a curious pseudo-vesicular
appearance to the weathered surfaces, as may be noticed on
Mynydd-y-Rhiw near Sarn and in other places. Only occasionally,
e.g. to the north of Pwllheli, do the diabases become visibly porphyritic. In a few localities there is a local schistose structure
due to shearing and internal slipping in the solid rock, as at
Rhyd-ddu and some spots near Tremadoc.

The jointing is rarely columnar (Clip-y-cilfinhir near Sarn[3]);
sometimes platy (Careg-y-rimbill at Pwllheli[4]); rarely cylindrical
(Craig-y-fael near Sarn); often roughly cuboidal or quite irregular.
A very common feature, especially in the Lleyn diabases, is a
pronounced spheroidal jointing, accompanied usually by shelly
exfoliation. This may be studied in endless variety near Sarn,
at Ty-rutten, in all parts of the Mynydd-y-Rhiw mass, and in the
sheet at Tyn-y-rhedyn near Llanfaelrhys. The spheroids vary
in size from two or three inches to as many feet in diameter.
In some localities, as on the beach west of the Mynydd Penarfynydd headland, they occupy the blocks marked out by three
sets of plane joints; but this is not always the case, and it is
not uncommon to find one large spheroid enclosing a number of
others like eggs in a nest (Coch-y-foel near Sarn).

Wherever the diabases are seen in conjunction with sedimentary strata, the contact-alteration of the latter is observable;
but it varies very much in degree and in the distance to which
it extends from the actual junction. In the field, indurated slates
or shales often assume a rather arenaceous appearance, and have
sometimes been erroneously mapped as sandstones, as is the case

---

[1] *G. M.*, 1888, pp. 303—308.
[2] *Q. J. G. S.*, vol. XLIV., pp. 448—450; 1888.
[3] Cf. Blake, *Q. J. G. S.*, vol. XLIV., p. 531; 1888.
[4] Cf. Bonney, *Q. J. G. S.*, vol. XXXII., p. 145; 1876.

in the Tremadoc district. One effect of the alteration has been to prevent the impression of the cleavage-structure on the argillaceous strata. Rocks which are strongly cleaved at a short distance from the intrusion are found to have resisted this modification completely in the immediate neighbourhood of the diabase. This phenomenon is rather different from that seen in the Llanberis slate-zone, when the cleavage has been obliterated by the effects of the intrusion of later (post-Carboniferous) dykes. It is well seen in Bwlch-y-ddeufaen and at other localities between Conwy and Snowdon.

A banded appearance is sometimes set up in the baked shales by the development of some alternation of lithological characters not perceptible in the unaltered rock. Some of the best examples of contact-metamorphism in Caernarvonshire are seen in the Tremadoc district, especially under the diabase outlier on Moel-y-gest. Here indistinct darkish spots make their appearance in the gray slates, which, as we approach the diabase, become smaller but more distinct. Still nearer, these spots, presumably incipient crystals of andalusite, seem to be reabsorbed and disappear, the rock becoming porcellanous with a subconchoidal fracture. Mr Teall (*Brit. Petr.* pp. 217—220) has compared altered rocks from this neighbourhood with the *spilosite* and *desmoisite* of continental geologists.

As a rule, each sheet or mass of diabase is fairly uniform to the eye ; but some shew segregation- or contemporaneous veins, which represent the latest phase in the process of consolidation. The general characters of these veins are, that they are more felspathic and so of lighter colour than the rest of the rock ; they are of coarser grain ; and they frequently have the augite partly idiomorphic, while in the matrix it is 'intersertal' or usually ophitic. Such veins sometimes occur pretty numerously, and have a roughly parallel or subparallel arrangement. Good examples are seen at Moel-y-gest and Pant-evan near Tremadoc, and in the boss known as the 'Gimlet Rock,' Careg-y-rimbill, at Pwllheli.

Examined microscopically[1], the diabases exhibit as a rule only the essential constituents, being decidedly poor in accessories. Any clear trace of olivine is almost unknown. Mr Teall (p. 215)

[1] Sixty-five slides have been studied.

has remarked the absence of olivine as one point of distinction between these rocks and those associated with the Carboniferous and Tertiary strata of Britain; but of the diabases and dolerites which may be regarded as post-Carboniferous in this part of the county only a small part contain olivine, and a separation on petrological grounds between the two sets of rocks becomes often impossible.

The original minerals composing the Caernarvonshire diabases are the usual iron-ores, triclinic felspars, and augite, with apatite in small quantity, possibly olivine in rare examples, and in one or two cases hornblende, biotite, and perhaps quartz.

A very characteristic feature in these rocks is the common association of magnetite and ilmenite. The original magnetite usually occurs in grains shewing little regularity of outline, in rods, or in tree-like growths produced by a series of minute rods set at right angles to a larger one which serves as a stem [580, 599, 733]. In several slides the magnetite rods, or perhaps sections of plates, shew a parallel arrangement over a considerable area [732], although the interspaces are filled by later minerals, and often each bar of the series is interrupted by a preëxisting crystal of felspar. This peculiarity of orientation, though found in the magnetite of some other rocks[1], is apparently not met with in other minerals. It is certainly not connected with fluxional movement, and may possibly be a magnetic rather than a crystalline effect. Occasionally magnetite rods appear to be built up from a series of imperfect octahedral crystals. More rarely this mineral, apparently titaniferous or intergrown with ilmenite, forms good skeleton octahedra like that figured by Zirkel from re-fused Mount-Sorrel granite [811].

The ilmenite commonly forms ragged plates, often arranged in parallel sets in a kind of frame-work. A section often shews two or three intersecting sets of parallel bars [731, etc.]. One or other of the two iron-ores is invariably present, and very frequently both may be recognised in the same slide. They do not entirely separate out at one stage of the consolidation, but in most cases the bulk of these minerals is later than the felspars

[1] Reusch, *The Microscopic Texture of Basalts from Jan Mayen* (from *The Norwegian North-Atlantic Expedition*, 1876—78), p. 5 and fig. 4; 1882. Judd, *Q. J. G. S.*, vol. XLII., pl. vi. fig. 7; 1886. See also *G. M.*, 1887, p. 413.

and earlier than the augite. We do not refer here to the secondary magnetite which often results from the alteration or destruction of the augite.

Apatite occurs in slender needles in many of the diabases, but only locally and capriciously. It cannot be reckoned as an essential component.

Sphene, excluding the amorphous leucoxene, has been found in one locality only, at Pant-Evan, Tremadoc. It forms clusters of light-brown grains with high refractive index. By comparison of different slides [576, 577, 578] this mineral appears to have been produced by reaction between the ilmenite and felspar. If this be so, it cannot properly be ranked among the original minerals, being probably connected with the mechanical forces which have deeply affected the rocks of that neighbourhood.

The felspar of the Caernarvonshire diabases belongs always to triclinic species, and appears to range from andesine to anorthite, varieties of the andesine-labradorite series being the most usual. In the majority of cases the felspar clearly antedates the augite in the process of consolidation, and builds idiomorphic, though not always well-bounded, crystals from 0·05 to 0·5 inch in length. The crystals are commonly of rather elongated form, but sometimes there are a few of squarer shape. These last often shew pericline twinning in addition to that on the albite plan, but most of the felspars have only the common albite-lamellation. In certain rocks Carlsbad twinning modifies the appearance of the striation in polarised light. Not infrequently some of the lamellæ project beyond the others, producing a ragged appearance at the terminations. Only in a few cases is there any suggestion of the twinning being a secondary structure related to strains in the crystals. In many of the rocks the smaller felspars are only once twinned. A zonary banding is sometimes seen in the felspars, when examined in polarised light.

Augite is invariably a plentiful constituent in the diabases. Only rarely does it occur in idiomorphic crystals, shewing in cross-section the prism and pinacoid forms about equally developed, as at Llanrwst [566] and at Pen-y-dre near Aberdaron [598], or the pinacoids only slightly truncated by the prism, as in some diabases north-west of Pwllheli [602], or again rather rounded outlines, as at the top of Carnedd Llewellyn [563]. Sometimes the augite is *partly* idiomorphic, presenting crystal contours to the felspar on one side but moulding it on the other, as near Llyn Teyrn [569], Sarn [731], and in the Pen-y-cil sheet near Aberdaron [594, 599]. Most usually the augite is entirely posterior to the felspar, and forms either grains filling the remaining spaces or sub-ophitic or ophitic plates moulding and enclosing the felspar crystals. Twinning on the orthopinacoid is not infrequent, and sometimes twin lamellation is seen [568, 580, 602].

The colour of the augite in thin sections varies from light brown to colourless, and is usually very pale, especially in the Lleyn diabases. Occasionally a crystal or plate of augite is separated by definite lines into fields of slightly differing optical characters. Such lines may be quite irregular, or may correspond to crystal outlines, as at Deneio near Pwllheli [603]. Sometimes they give rise to a well-developed hour-glass structure. In certain cases this is seen only in polarised light, when a clinopinacoidal section may shew a difference of perhaps 5° between the extinction-angles for the two parts of the crystal, as at Castell Caeron and other places near Sarn [580, 811]. In other specimens there is a difference in natural light between the augite-substance within and without the 'hour-glass,' the former being colourless, while the latter is brown and more or less pleochroic. This is seen at Pen-y-dre, near Aberdaron [598], Deneio, near Pwllheli [603], and Porth-lleiddiad, near Llanfaelrhys [734]. It is to be noticed that the hour-glass structure is rarely seen except in idiomorphic crystals, and that the external part is always more coloured, presumably richer in iron, than the internal. It may be suggested with some reason, that the structure is produced by a crystal being corroded by magmatic resorption, and subsequently built up by an accretion of new substance from the magma, which has become altered in composition by the resorption of some of the earlier iron-ores. The interesting synthetic studies of Famintzin[1] have shewn that an hour-glass form is readily produced in crystals corroded by their mother-liquor.

The augite of the Caernarvonshire diabases always shews good prismatic cleavage, and rarely others parallel to the two pinacoids,

---

[1] *Studien über Krystalle und Krystallite*, Mém. Acad. St. Petersb., vol. xxxii., no. 10; 1884: see plate i., figs. 14, 15.

as near Pen-y-cil, and Deneio, and in a sheet near Sarn [594, 602, 731]. Very rarely is there any diallagic structure, and that only locally [Deneio, 603]. In one type of diabase the augite is traversed by numerous roughly parallel cracks and rows of thickly-set minute inclusions. This is seen in the rocks of Careg-y-rimbill, near Pwllheli, Llanengan, near Abersoch, Careg-dinas, north of Carn Fadryn, and Trefgraig, four miles west of Sarn, rocks which are also united by other characters [136, 582, 135, 592].

The ordinary decomposition-products of the augite will be noted below. The augite in the diabases very rarely shews any trace of bodily conversion into hornblende; but a fringing growth of amphibole bordering the augite, traversing it in strings, and extending into the place of decomposed felspar and possibly olivine, is an exceedingly common feature in the diabases of the eastern half of Caernarvonshire. This secondary amphibole is always in crystalline relation with the augite on which it grows. It is colourless or rarely pale green[1], and has commonly a fibrous structure [566, 568, 570, 571, 573, 687]. The Lleyn diabases never have this 'secondary enlargement' of augite by hornblende. At Pant-evan, near Tremadoc, a pale green amphibole with similar characters and relations forms not only fibrous fringes, but round compact patches with hornblende-cleavage, partly filling cavities in the augite which also contain serpentine [576].

A few scraps of brown hornblende and brown biotite associated with the augite may probably be reckoned as original. They are seen occasionally in the Careg-y-rimbill type of diabase already alluded to, with which perhaps we may include the occurrences near Tremadoc [136, 583, 135, 577]. Only one slide, from Llyn-yr-afon under Y Foel Frâs [855], shews a few small idiomorphic crystals of pale brown hornblende, of earlier formation than the augite.

The general structure of the rocks shews certain varieties, but as a rule the whole of the felspar has been of earlier consolidation than the augite, and the various types of 'diabasic' structure result. There is every gradation, from large ophitic plates of augite including many felspar crystals [567, 136, 576, 580] to small ophitic patches just moulding the felspars [602, 731], and augite grains wedged into the interspaces of the earlier-formed

---

[1] Very rarely brown [855].

minerals [728, 600, 589]; but the ophitic structure is more common than the so-called 'granulitic.' Any approach to the gabbro-type is rare, but there are some diabases in which the felspar and augite seem to have crystallised concurrently, and both have more or less idiomorphic contours, as at Llyn Teyrn, and in some sheets near Sarn [769, 731]. Again, the augite may occur in shapeless grains, and yet penetrate, as well as being penetrated by the felspar prisms, as between Pen-y-cil and Aberdaron [599]. A rare variety is one near Llanrwst, in which the augite occurs in two distinct generations, *viz.* in long twinned prisms older than the felspar and in ophitic plates moulding that mineral [566]. One or two specimens come near to the basaltic type: a rock from the summit of Carnedd Llewellyn, for example, has rather rounded crystals of twinned augite, with felspars in smaller number, imbedded in a fine-grained ground which is now much decomposed [563].

The numerous decomposition-products met with in the weathered diabases of Caernarvonshire will be only briefly noticed. The ilmenite and perhaps titaniferous magnetite give rise invariably to the usual leucoxene, clinging in cloudy semi-opaque masses to the remains of the iron ore. In some of the diabases from the eastern half of the county the leucoxene becomes light-brown and slightly doubly refracting, but distinctly granular sphene has not been found except in the Pant-evan rock, as remarked above. The pale secondary amphibole already described seems to come less from the augite than from the destruction of the other constituents. A fibrous pilitic-looking substance forms felted masses in some of the round pseudomorphs which may represent olivine, *e.g.* on Moel Hebog [565]. A bastite-like material at Pant-evan, *etc.*, may possibly come from a rhombic pyroxene, and, if so, the association of the minerals would indicate the same source for the *compact* variety of the pale secondary amphibole, which would thus seem to pseudomorph an enstatite mineral originally intergrown in the augite plates [576].

The pale green decomposition-products so prominent in many of the slides are doubtless to be referred to more than one mineral of the chlorite and chloritoid (or saponite) families, besides a little serpentine. Professor Heddle[1], who has devoted much attention

---

[1] *Trans. Roy. Soc. Edinb.*, vol. xxix., pp. 55—118; 1879.

to the 'chloritic minerals,' divides them into the 'chlorites' and the 'saponites,' and makes it appear that the ordinary decomposition-products of augite, hornblende, etc., belong to the latter division. The four chief species of this division are celadonite (bright green), delessite (dark green), chlorophæite (usually black), and saponite (translucent). The first two have a comparatively low content of water, with a high specific gravity; the last two a high percentage of water and a low specific gravity. All are dissolved or destroyed by hydrochloric acid except celadonite, which in this respect resembles the chlorites proper. The substances met with in the Welsh diabases seem, from the criteria here cited to be delessite, and perhaps sometimes chlorophæite. Unfortunately the optical properties of the several species are not thoroughly known, and indeed small reliance can be placed on statements, unless chemical and optical examinations have been made on the same specimens. The common 'viridite' of our diabases has usually a low refractive index and very slight double-refraction, with varying degrees of colour and pleochroism in thin sections. These, according to Rosenbusch[1], are the characters of delessite. Sometimes, however, and especially in the radial-fibrous aggregates, the double refraction is much more marked. The vermicular structure is rare [Pant-evan, 578].

These green decomposition-products often form evident pseudo-morphs after the augite, especially after the ophitic plates; but sometimes a sensibly isotropic substance, probably the same, impregnates the rock generally, and is even injected into the cleavage-cracks of the felspar, as in the Pen-y-cil sheet near Aberdaron [594]. Doubtless the production of this substance from the augite was attended by an increase of volume.

Serpentine occurs in many of the slides, either mingled with the presumed delessite or alone. In the Deneio rock are numerous reticulating veins of fibrous chrysotile, the fibres set perpendicular to the walls of the fissure [603]. The rock tends to break under the hammer along these serpentinous veins, shewing a beautiful velvet-like surface. Magnetite dust is not infrequently produced by the alteration of the augite in all the diabases, and is seen either in the augite or among its decomposition-products. The

---

[1] *Mikr. Phys. d. petr. wichtig. Miner.*, p. 560 (2nd ed.), 1885.

separation of this dust seems to be quite an early stage in the decomposition of the augite.

Epidote in brightly polarizing grains is a frequent secondary product in the diabases of Caernarvonshire[1], although restricted to the eastern half of the county. It appears to be connected with the felspar rather than the augite, and sometimes builds more or less perfect pseudomorphs after the felspar crystals [568— 571 and 856]. In other cases the distribution of the grains is less regular; and when the epidote is present in large quantity, it forms large fan-like bundles of prisms. Calcite is common in dust or fine granules: only in the more weathered rocks does it build larger grains, and it is then sometimes accompanied by secondary quartz with fluid-cavities enclosing moving bubbles. Some of the diabases of Deneio and a few other localities have fan-shaped aggregates of a zeolitic mineral with high polarization-tints [602]. Among clearly introduced constituents may be mentioned pyrites, sometimes in small cubic crystals, as at Tremadoc and at Pen-y-dre, near Aberdaron.

In a few cases only has the diabase undergone so much alteration as to be completely disintegrated. This is the case north of Coch-y-foel, near Sarn, where the rock is converted into a soft brown ochreous substance.

In the extreme west of the Lleyn peninsula occur certain basic rocks of volcanic *habitus,* which are too much decomposed to admit of profitable examination, but which we may fairly refer to the basaltic family. They are associated with the doubtful series of green schists, *etc.*, which stretch in a broad band from Porth-dinlleyn, near Nevin, to Braichy-y-pwll and Bardsey. This remarkable series of rocks, the 'altered Cambrian' of the Geological Surveyors and the 'Pebidian' of Dr Hicks, is frequently schistose and often closely contorted; usually green from chloritic substances; in some places quartzose, in others felspathic, occasionally micaceous, often including streaks and nodules of red jasper. It not infrequently includes calcareous beds, and even more or less

---

[1] *Cf.* Cole and Jennings on the Cader Idris diabases, which, however, appear to be less basic than those under discussion. *Q. J. G. S.*, vol. xlv., p. 432; 1889.

crystalline limestones and ophicalcites, and is intersected by count-
less dykes well exposed on the sea-shore. The stratigraphical
position of these rocks cannot be discussed here. They exhibit,
however, the closest resemblance to those exposed in various parts
of Anglesey, particularly near Menai Bridge, Holyhead, and Am-
lwch. These Anglesey strata Professor Hughes[1] has proposed to
refer, at least in part, to the Bala series; and Professor Blake,
while placing them beneath the Cambrian system, admits their
continuity with the conglomerates at Porth-wen on the north
coast, which have yielded *Orthis Bailyana!*

Within the tract mainly occupied by these rocks in western
Lleyn a number of small patches are coloured on the Survey maps
as 'serpentine.' As remarked by Professor Bonney[2], Mr Elsden,
and Professor Blake[3], they have no real claim to this title. In
part they are diabases, and have been included among the rocks
described above; while the remainder are much decomposed
basalts. The 'serpentines' of Methlan, Ty-hen, Hendrefor, and
Trefgraig [592] are diabases; those of Porth-dinlleyn and Careg-
fawr are basalts, although a diabasic rock is associated with the
occurrence in the former locality.

In appearance these weathered basalts are compact, brittle
rocks of green to purplish red colour, veined with calcite, quartz,
epidote, and sometimes asbestos and chrysotile. In the Careg-
fawr patch near Aberdaron there are abundant veins lined with
epidote and containing asbestos: the former mineral grows perpendicular, the latter parallel to the walls of the fissure.

In thin sections from either locality [591, 612, 613] we see a
decomposed mass, in which small felspar prisms and sometimes
augite granules are still recognisable. There appears to have
been an isotropic base, and small curving cracks, occasionally
outlined in secondary magnetite dust, may represent a perlitic
structure. Larger cracks, producing a brecciated appearance in
the slides, are filled either with calcite or with a mixture of
serpentine and a chloritoid; and it is the fracture of hand-specimens along these cracks which gives a false serpentinous aspect
to the rock in the field. The deception is heightened by the

[1] *Proc. Camb. Phil. Soc.*, vol. iii., pp. 341—348; 1880.
[2] *Q. J. G. S.*, vol. xxxvii., p. 48; 1881.
[3] *Q. J. G. S.*, vol. xliv., p. 531; 1888.

mottled green and red appearance of the rock, due to some capricious oxidation of the contained iron-compounds.

Professor Bonney described the basalt at Porth-dinlleyn as associated with volcanic agglomerate and breccia[1]; but he informs me that he did not regard this as evidence that the volcanic rocks were contemporaneous with the green schist group, the appearances being more suggestive of a 'neck,' like those of Fifeshire. The basalt at Careg-fawr, where the relations are more evident, is clearly intruded through the surrounding rocks. The age of these latter being so dubious, we cannot pretend to fix the date of the basalts, and the above notice of the existence of such rocks in a couple of small exposures will suffice.

---

[1] See also Harrison, l. c.

AMONG the smaller eruptive masses of Lleyn, two intrusions of basic rocks, breaking through the granite of the Sarn district, are deserving of notice[1]. The localities are Craig-y-fael, two miles south-west of Sarn, and Plas Llangwnadl at the same distance west-by-north of that village, the rocks being quarried in both places. (See map, fig. 4.)

In the Craig-y-fael quarries the rock exposed shews a gabbro-like structure, and appears to consist of darkish grass-green bisilicates with a rather sub-metallic or diallagic lustre and light-green altered felspars. These two constituents are present in about equal proportions, and neither shews any idiomorphic contour. At the bridge above Plas Llangwnadl, on the main-road from Nevin to Aberdaron, the exposures are very similar, though sometimes with apparently less felspar in the rock. Tracing the mass down the stream, past the Plas, the character seems to vary, with some tendency to a gneissose structure. One variety with large white felspars has to the eye a suggestion of the *augen*-structure. Going on towards Llangwnadl itself, the rocks become distinctly schistose ; and about a hundred yards before reaching the church they suddenly disappear, doubtless against the same fault which limits the granite in this direction, being succeeded by schistose, agglomeratic-looking rocks belonging apparently to the green schist series, and shewing evidence of fault-brecciation.

The Craig-y-fael type may be described as a partially am-phibolised gabbro. The minerals seen in a thin section are augite and diallage, hornblende and actinolite, felspar and opaque

---

[1] See *Q. J. G. S.*, vol. XLIV., pp. 447, 448; 1888.

iron-ore, with some pale-green decomposition-product which has
the characters of a chloritoid [706].

The felspar is a triclinic one with the usual albite-twinning,
sometimes crossed by pericline lamellæ, and the extinction angles
seem to point to labradorite. The crystalline plates are much
strained and bent, the lamellæ being curved. It is possible,
though not quite clear from the specimens, that the twinning
may be in part induced by the strain.

The augite is in long plates, which exhibit sometimes the
ordinary prismatic cleavage, sometimes a distinct diallagic structure. They have a slight greenish tint, and, apart from the
ordinary decomposition, are seen to pass at the margin and in
irregular veins and patches into a dull green hornblende. The
latter mineral gives the absorption-formula

α.   very pale to almost colourless ;
β.   rather pale grass-green ;
γ.   a rather deeper and bluer green.

Some of this hornblende is uralitic in appearance, but there
are more compact portions shewing the prismatic cleavage; and,
in addition, granular patches and imperfect blades of actinolite are
met with.

There appears to be no augite or diallage remaining in the
Llangwnadl rocks, but their essential identity with the partially
amphibolised gabbro of Craig-y-fael can scarcely be doubted after
an examination of the specimens and slides. The alteration of
the original type has evidently proceeded farther near the margin
of the district than in the central part, and the resulting rocks are
such as would be termed epidiorites by some petrologists.

The massive type of these completely amphibolised rocks is
illustrated in the quarries near the bridge [707]. The microscope
shews but little of the black iron-ore, its place being occupied
by granules of brownish sphene. These granules, with cleavage-
traces and more irregular fissures, are aggregated in small patches
and strings between the grains of felspar and hornblende, but they
seem to be clearly connected with the diminishing iron-ore, and
may be taken to prove that the original mineral was a titaniferous
one. Possibly some part of the iron-oxide has been absorbed by
the hornblende, which here presents a greenish-brown to brownish-
green colour, with pleochroism :

α. very pale to almost colourless:

β. rather deep brown with a greenish tinge;

γ. a slightly deeper tint of greenish brown:

the absorption being, as before,

$$\gamma > \beta >> \alpha.$$

The felspar is partly in granular patches, but too much decomposed to shew any structure.

Near Plas Llangwnadl the rock is still more altered, and has in places the quasi-porphyritic aspect already noted, which is due to granular felspar forming large patches. There is but little opaque iron-ore, and sphene is more abundant. The felspar, where it has preserved something of its original structure, appears to be labradorite [708].

Still further down the stream, approaching the boundary of the area, the schistose structure becomes more pronounced, and the rock resembles the 'amphibolites' of some Continental geologists. A slide from this portion shews no new characters, except that the iron-ore has entirely disappeared [709]. Again, a gneissic structure is sometimes to be seen, and this rock is illustrated by another slide, the material, however, not obtained in place [710].

The mode of occurrence of the rocks in both the small masses proves them to have been intruded through the granite, doubtless consolidating in the form of gabbro. The various mineralogical changes must be referred to mechanical influences, and indeed are well-known results of dynamic metamorphism. The manner in which these modifications increase towards the boundary of the northerly intrusion, and are increasingly attended by structural re-arrangement of the mass, clearly point to a relation between these changes and the forces which produced the Llangwnadl fault. The strike of the schistose structure in the amphibolite-like and gneissic rocks is parallel to the fault, as is also the local schistosity in the rocks on the other side of the dislocation.

As regards the age of the intrusions, it can only be said that they are posterior to the granite and anterior to the fault, which leaves the upper limit of age rather indefinite. No post-Silurian or post-Carboniferous crust-movements appear to have been accompanied in North Wales by structural metamorphism, and so the gabbros are most probably assigned to the Bala age, which we know to have been closed by extensive movements of the earth's crust in the Caernarvonshire area.

Professor Blake[1] includes the rocks in question with the granite of Cefn-amwlch, *etc.*, but has withdrawn his former opinion assigning the whole to his 'Monian' system. The gabbro and its derivative rocks are, however, as the above description indicates, thoroughly basic rocks, while the granite is always highly acid; and there is no appearance of passage between the two.

In the southern part of the Sarn district, an area remarkable for the variety of its petrological types, occurs a considerable development of coarsely crystalline hornblendic rocks, differing entirely from the neighbouring diabase. I have elsewhere described these rocks under the title hornblende-diabase, and given some account of their mode of occurrence[2]. They were first noticed by Mr Tawney[3], and Mr Elsden[4] has also remarked upon them. They occur in three apparently distinct masses, each having a roughly oval outline on the map (fig. 4), with its long axis bearing N.N.E.-S.S.W., that is parallel to the local strike of the strata. The mode of occurrence of these masses and the contact-alteration of the adjacent shales attest their truly intrusive nature. The first forms the hill Mynydd Penarfynydd, and has a length of about three-quarters of a mile. The second builds Careg-llefain and Mynydd-y-graig, being rather larger than the preceding. The third, less freely exposed, lies to the east of Rhiw, and extends from Tyn-y-borth to Tyddyn-y-corn, a distance of more than a mile.

The Mynydd Penarfynydd mass is underlain by a considerable thickness of hornblende-picrite, exposed on the west and southwest flanks of the hill, which is some 360 feet in height, and the whole, with a small thickness of hornblende-diabase at the bottom,

---

[1] *Q. J. G. S.*, vol. XLIV., pp. 531, 532; 1888.
[2] *Q. J. G. S.*, vol. XLIV., pp. 450—459; 1888.
[3] *G. M.* 1880, pp. 207—213.
[4] *G. M.* 1888, p. 305.

rests on shales, in which Mr Tawney[1] found Upper Arenig fossils at Penarfynydd farm. A comparison of the cliff-sections with the outcrops in the hill shews that the intrusion is to be regarded as a laccolite, probably more than 1000 feet thick, injected between the strata. The second mass rises to more than 700 feet in the serrate ridge of Mynydd-y-graig, but unfortunately neither this intrusion nor the one near Rhiw, which occupies lower ground, shews any clear junction with the adjacent rocks. They may be laccolites of larger size than that of Penarfynydd and injected at a rather higher horizon. The manner in which all these rocks have modified the folding and cleavage of the district conclusively proves them to be of Bala age.

In hand-specimens the rocks are always visibly crystalline, of medium to rather coarse grain, and shew a mixture of hornblende and felspar; but the variation in colour and general appearance is so great that no single description will serve for the whole of the rocks. As in some other hornblende-diabases, and notably those of central Anglesey, the diversity of appearance is due to the varieties and varying relations of the hornblende rather than to any essential differences in constitution among the rocks themselves. The structure is normally 'hypidiomorphic', though in one or two localities (north of Rhiw Church) the rock contains more or less well-formed crystals of hornblende some two inches in length.

Mr Acton's analysis shews 44·9 *per cent.* of silica, thus placing the rock naturally between the ordinary diabases (47·4 *per cent.*) and the hornblende-picrite (41·8). In many respects the hornblende-diabase is allied to the latter, but differs from it mineralogically in the absence of olivine and the presence of plenty of felspar. There is a considerable quantity of alkalies, the percentage of potash being 2·1, and of soda 3·6. A loss of 1·5 *per cent.* on ignition at red heat represents the water in the serpentinous and chloritoid decomposition-products.

An examination of slides of the hornblende-diabase from various localities shews, among the original minerals, apatite, magnetite, picotite, ilmenite, felspar, olivine, augite, and hornblende.

Apatite occurs only rarely in rather large prisms of a pinkish

[1] *G. M.* 1880, p. 211.

tint with transverse jointing (east flank of Mynydd Penarfynydd [713]). Magnetite, in grains of primary origin, is frequent, and sometimes abundant. One specimen (west of Y Graig-ddu [716]) has deep brown rounded grains of picotite; another (north of Rhiw [719]) has ilmenite in skeletons of intersecting rods, with grey leucoxene; but in the other slides magnetite is the only iron-ore present. The above minerals are the first products of con-solidation.

Olivine is not actually found in any of the slides, but in one or two some serpentinous pseudomorphs included in the hornblende seem to represent that mineral. The specimens are from near Ffynnon-cefn, north of Rhiw [718], and from the top of Mynydd-y-graig [715].

Augite occurs sometimes in idiomorphic crystals, but usually in ophitic plates of varying extent or in shapeless crystalline grains. The forms seen in cross-sections are, as usual, the pinacoids truncated by the prism [718, 712]. The prismatic cleavage is constant, pinacoidal cleavages rare; while the diallagic structure is never found. Twinning on the orthopinacoid is not infrequent. The colour of the mineral is very pale brown to colourless: in one specimen from the top of Mynydd-y-graig [715] this passes into violet-brown with slight pleochroism. The hour-glass structure is occasionally met with, as near Ffynnon-cefn [718]. The extinction-angle in a clinopinacoidal section is 39° or 40°.

Hornblende in these rocks is both an original and a se-condary mineral. It very rarely shews idiomorphic outlines, viz. the prism and clinopinacoid without terminal planes [719]. The mineral almost always occurs in ophitic plates or in patches or borders associated with the augite. Twinning is rarely seen: it is on the common law, the orthopinacoid being twin-plane and plane of composition. Whether primary or secondary, the horn-blende is normally compact and shews the prismatic cleavage well marked: a fibrous structure is only occasionally met with. Both original and derivative hornblende are usually of a fine brown colour, giving

α, deep chestnut brown;

β, a less deep tone;

γ, pale brown:

absorption $\gamma \geqq \beta >> \alpha : \beta$ coincides with b, and the extinction-angle $c\gamma$ is a rather large one, measuring 18° or 20°. The colour sometimes passes into greenish-brown or green.

As regards its mode of growth and relation to the augite, the ordinary brown hornblende shews some interesting differences. A careful examination of the sections enables us to discriminate four different cases[1].

(i) Original idiomorphic hornblende, which is only rarely met with.

(ii) 'Complementary' hornblende, forming a border to original augite, the marginal hornblende of each plate or crystal being in crystallographic relation to the augite nucleus; *i.e.* the two minerals have the vertical axis and the plane of symmetry common. This must be regarded as an original 'intergrowth' of augite and hornblende, in which, however, the hornblende is entirely posterior to the augite [716]. This relation is not uncommon in hornblende-diabases (proterobases) from other localities. and is most beautifully exhibited in one from Delancy Hill, Guernsey [431]. In the rocks under discussion it is less obvious, but very common. In one case [717] both hornblende and augite are twinned, and the twin-planes of margin and nucleus, though of course parallel, are not coincident, as is the case when the hornblende is pseudomorphic after augite.

(iii) Ophitic hornblende, which forms plates moulding and including the felspars, magnetite-granules, and grains of augite. Sometimes even well-formed augite crystals are thus enveloped by hornblende, as at the trigonometrical station on Mynydd Penarfynydd [712]. When augite grains are included, their boundaries are comparatively smooth and rounded, and those within the same plate of hornblende have different orientations, there being no crystallographic relation between the augite and the investing hornblende.

(iv) Pseudomorphic ('paramorphic') hornblende, formed at the expense of augite, portions of which often remain unaltered. Here, as in the complementary hornblende, the two minerals have the vertical axis (c) and the plane of symmetry in common; but the line of demarcation is an exceedingly irregular and intricate one. The change in the augite begins sometimes at the margin,

---

[1] *Cf.* also *Min. Mag.*, vol. VIII., pp. 30—33; 1888.

sometimes along cleavage-planes, sometimes along planes parallel
to the pinacoids, or again in quite capricious patches. In some of
the slides every stage of the process can be studied, from grains or
idiomorphic crystals of augite with small patches of hornblende,
mostly near the margin, to plates of hornblende enclosing mere
shreds of augite. Here all the patches of augite within one hornblende plate have, of course, one orientation, being relics of one
original crystal.

This pseudomorphic hornblende is doubtless for the most part
truly secondary in the ordinary sense of the word. In one or two
examples, on the other hand, there are appearances which suggest
that the amphibolisation may have begun before the final solidifi-
cation of the rock; as, *e.g.*, when a grain of augite is seen partly
pseudomorphed by hornblende which is in crystalline continuity
with hornblende moulding the grain [714].

The pseudomorphic hornblende in these rocks, though sometimes of a rather paler brown, is usually indistinguishable in
colour and structure from the original hornblende in the same
slides. Another kind of secondary hornblende has different
characters, and is probably of later formation. It is seen as a
colourless or green extension in crystalline continuity with older
hornblende, but occupying the place of other minerals which
have perished. This is seen in a number of sections [712, 717,
718]; but the secondary fibrous amphibole fringing plates of
augite, which is so constant a feature in the diabases of eastern
Caernarvonshire is absent in these hornblende-diabases.

The brown hornblende occasionally passes over into uralite, as
on the south-east slopes of Mynydd-y-graig, near Y Graig-ddu
[716], or into actinolite, as on the east flank of Mynydd Penarfynydd [713].

Felspar is always an abundant constituent of the hornblende-
diabases, though often much affected by decomposition. It belongs
to basic varieties between labradorite and anorthite, or to the
latter species, as shewn by the wide extinction-angles [717]. The
crystals are sometimes simple, generally twinned once or twice,
and in some cases finely lamellated: pericline twinning is, however, very rarely seen. In almost every case the felspar is of
earlier consolidation than the bisilicates, which frequently mould
it in ophitic fashion.

The decomposition-products of these rocks call for no special notice. The augite and hornblende often pass into a substance of the chloritoid family (delessite?), but both minerals also give rise to serpentine. The augite is also converted into hornblende, as described above, while in some cases fibrous uralite, blade-like actinolite, or a dark micaceous mineral result from the alteration of the compact brown hornblende. Calcite dust usually accompanies the decomposition of the felspars, and secondary quartz occurs in a few cases. One or two slides [719] have fan-like bundles of a zeolitic mineral polarizing in colours of a high order. The separation of magnetite dust often marks the first stage of alteration in the brown hornblende.

It will be seen that these hornblendic rocks belong to a type widely different from the ordinary diabases of the county, which do not contain original hornblende. A rather different type of hornblende-diabase, with olivine, will be described below, but we shall first discuss the ultra-basic rocks, also hornblendic, which occur in close association with the Penarfynydd hornblende-diabase.

The hornblende-picrite of Mynydd Penarfynydd forms the basal portion of the laccolite already mentioned. The rock occurs in a succession of parallel banks or quasi-strata to a total thickness of 200 or 250 feet. Owing to a steady dip of 40° to S. 30° E., it is exposed only on the west and south-west slopes of the hill, being succeeded above by the hornblende-diabase which builds the bulk of the laccolite, while there is a thin band of a rather different hornblende-diabase [711] below, in contact with the sub-jacent Arenig shales[1]. (See map, fig. 4.)

The rock may be regarded as the type of hornblende-picrite. It was first described by Professor Bonney[2], and studied in the field by Mr Tawney[3], who was doubtfully of opinion that the rock is intrusive in the neighbouring hornblende-diabase. An examination of their junction below the trigonometrical station on

[1] For further description, see *Q. J. G. S.*, vol. xliv., pp. 454—459; 1888.
[2] *G. M.*, 1880, p. 208: *Q. J. G. S.*, vol. xli., p. 517; 1885.
[3] *G. M.*, 1880, pp. 208—211.

Mynydd Penarfynydd shews, however, that although there is no very gradual transition between the two rocks, they are evidently in close connection, and indeed have segregation-veins passing from one to the other. These veins are of coarser grain than the normal type of the rock, and more felspathic: unlike either of the normal rocks, they frequently contain well-formed crystals of hornblende with terminal faces: they have no olivine. Similar veins occur in the heart of the hornblende-picrite itself, and have usually a direction roughly parallel to the quasi-strata already alluded to. They are doubtless in a general sense contemporaneous veins, though representing the latest phase in the consolidation of the magma. They are most abundant in the upper part of the picrite.

In the field and in hand-specimens the hornblende-picrite has a very striking appearance. The most prominent mineral is hornblende, the lustrous cleavage-planes of which are seen to be studded with dull round spots representing grains of partially serpentinised olivine. One common variety contains a golden-brown mica, the flakes lying mostly along the cleavage-planes of the hornblende. White felspar crystals are abundant in some parts of the rock, but wanting or almost wanting in others. (It will be remembered that Tschermak's[1] picrite contained felspar up to about 25 *per cent.*) The several varieties form clearly marked banks like beds, all parallel to the plane base of the laccolite and to the stratification of the shales beneath. The bedded aspect is very remarkable in the field, being strongly brought out by differential weathering. There are two main types of structure: the first, most noticeable in the felspathic variety of the rock, exhibits a partial separation into patches of the component minerals, which produces a mottled or spotted appearance on smooth faces, as in the boulders on the beach, and a pitted or honey-combed effect on a weathered surface. The second type, seen when the constituents are more evenly distributed, shews a compact appearance, with usually a smoother weathered surface, though there is often a grooved or fluted aspect, due to the alternation of bands rich and poor in olivine, and even thin seams composed mainly of the decomposition-products of that

---

[1] *Sitz. k. k. Akad. Wien*, vol. LIII., (1), p. 260; 1866.

mineral. Somewhat similar alternations of more and less basic layers of rock are described by Reusch in the 'saussurite-gabbros' of the Bergen district[1]. In Mynydd Penarfynydd the mottled or honeycombed type of structure is seen chiefly in the middle half of the mass, though it contains bands and layers of the compact and fluted varieties.

Mr Acton has kindly analysed the hornblende-picrite (a felspathic variety [725]). Its ultra-basic nature is shewn by the silica-percentage, 41·8, which corresponds very closely with Fuchs' analysis of the Schriesheim hornblende-picrite[2], (41·44 per cent.), a rock presenting close resemblance to the Caernarvonshire one. No titanic acid, chromium, or manganese was found, and only a trace of phosphoric acid. The percentages of the alkalies are, as might be expected, very low, viz. 0·2 of potash and 0·5 of soda: in the Schriesheim rock they are 0·93 and 0·24 respectively. Owing chiefly to the quantity of serpentine present, there is a loss of 3·6 per cent. on ignition at a red heat (Schriesheim 5·60). The Penarfynydd hornblende-picrite is rather more basic than that of Professor Bonney from Ty-croes in Anglesey, in which Mr J. A. Phillips[3] found 42·94 and 42·79 per cent. of silica.

The microscope reveals magnetite, olivine, felspar, augite, hornblende, and brown mica, besides serpentine, actinolite, asbestos, a chloritoid substance, calcite, dolomite, and other secondary products.

Here, as in many other basic and ultrabasic plutonic rocks, original magnetite is not common; but good crystals sometimes occur as the first-formed constituent [723]. As usual, granular secondary magnetite is abundant as an alteration-product of the olivine.

Olivine is always present in force, rarely presenting an idiomorphic outline, but commonly in rounded grains imbedded in augite or hornblende. It sometimes has minute, flat, rectangular cavities or 'negative crystals', containing dendritic growths of magnetite of the kind referred by Professor Judd[4] to secondary

---

[1] *Silurfossiler og pressede Konglomerater*, Christiania, 1882: (trans. *Fossilienführenden kryst. schiefer*, Leipzig, 1883).

[2] *Neu. Jahrb. f. Min. etc.*, 1864, p. 326.

[3] *Q. J. G. S.*, vol. xxxix., p. 256; 1883.

[4] *Q. J. G. S.*, vol. xli., p. 385 and pl. xii., fig. 5; 1885.

schillerisation [726]. The fissures which traverse the olivine-grains rarely follow any definite cleavage-direction. So far as I have noticed in this and other rocks, the most regular cleavage is found in those olivines which shew the most perfect crystal forms [721]. It is along the cracks that the process of serpentinisation, which may be observed in all its stages, first takes effect [720, 726, *etc.*]

Felspar, when present in the rock, stands next in the order of crystallization. It occurs either in small slender crystals, or in irregular plates moulding the olivine. Its extinction-angles agree with those of anorthite.

The augite is very pale brown or almost colourless, with pronounced prismatic cleavage. It shews in rare cases an octagonal cross-section, but is almost always in irregularly-shaped plates. The hornblende has the rich brown colour and the other optical characters already noticed in the hornblende-diabases. Occasionally it passes into a green variety giving for vibrations parallel to

α, very pale brown;

β, pale olive-green;

γ, rather pale grass-green.

It also passes into a colourless variety. The usual prismatic cleavage is in a few cases supplemented by one parallel to the clinopinacoid.

The hornblende never shews idiomorphic contours, and is usually in close connection with augite, which it includes or borders. In each plate the two minerals have the c-axis and the plane of symmetry in common, and a section parallel to this plane gives extinction-angles of 20° for the hornblende and 40° for the augite on the same side of the vertical axis. It seems probable that much of the hornblende is pseudomorphic after augite, the boundary between the two being commonly very intricate and ragged [720, 722, 725]; but it is also possible that there is some original intergrowth of the two minerals, as already remarked in the case of the hornblende-diabases. Again there is original hornblende which seems to be quite independent of augite [723].

The brown mica has the characters of biotite, including the usual absorption and dichroism, but it often becomes paler. It is in part clearly original, being then later than the felspar and earlier than the hornblende. Another portion, however, is produced at the expense of the hornblende, being sometimes parallel to the orthopinacoid, but more commonly lying on the prismatic cleavage-planes of that mineral.

The alteration of the hornblende-picrite presents no peculiar features. The olivine, quite fresh in many of the specimens, shews, when a number of slides are compared, the ordinary series of changes: first the separation of granular magnetite in strings following the cleavage or other less regular fissures; next the production of fibrous, unaxial serpentine on the walls of these fissures; then the gradual encroachment of the serpentine network upon the meshes of olivine, the fibrous structure being wanting in this later serpentine, which has a confused arrangement, and is often sensibly isotropic; and finally in many cases a resorption of the iron-oxide, resulting in a green coloration of the serpentinous pseudomorph[1]. The other constituents of the rock give rise to the products already noted in the case of the hornblende-diabases.

The comparison of the Mynydd Penarfynydd rock with other picrites has already been made by Professor Bonney. It resembles perhaps most closely the well-known type from Schriesheim near Heidelberg [169], although in the latter the hornblende is of a much paler colour and the mica quite bleached, while the olivine, in the specimens I have seen, is more decomposed. The rocks from Ty-croes in Anglesey and Little Knott in Cumberland [170] have less olivine than the one described above. In the St David's boulder [439], noticed and described by Professor Bonney, the olivine and augite are more decidedly idiomorphic than in our type. None of these seem to have such a variety of relation between the pyroxene and amphibole as the Penarfynydd specimens exhibit; but they shew many minor points in common, such as the secondary enlargement of the hornblende, the gradual passage of the brown variety into green or colourless, and the bleaching of the biotite. Of the Peekskill hornblende-picrite, 'hudsonite' of Cohen[2], and 'hornblende-peridotite' or 'cortlandtite' of Professor G. H. Williams[3], who has fully described its characters,

---

[1] *Cf.* Wadsworth, *Lithological Studies*, Cambridge (Mass.), 1884.
[2] *Neu. Jahrb. f. Min. etc.*, 1885, vol. II., p. 242.
[3] *Amer. Journ. Sci.* (3), vol. XXX., p. 29; 1886.

I have seen no specimens.   The same remark applies to the Gipps'-land picrite, the green bornblende of which is considered by Professor Bonney[1] to be largely secondary after pyroxene, and to the Cornish picrites.   The Inchcolm rock [820] seems to differ[2] in many respects from the Caernarvonshire one, and the Shrop-shire picrite[3] represents a widely divergent type.

A well-marked rock-type, though with some characteristic variations, is seen in the hornblendic rocks which occur in the neighbourhood of Clynog-fawr.   They form two intrusive masses. One builds a low ridge running for about 300 yards, in the direction of strike of the adjacent rocks, from the farm of Pen-y-rhiwiau[4] nearly to the cliff.   The other crosses the road eastward from Clynog-fawr, just east of the word 'Tanrallt' on the Survey map. Their intrusive nature is proved by the contact-alteration of the neighbouring slates, which, as is often the case, assume to the eye a more arenaceous aspect near the junction.   These slates are probably referable to the Arenig stage, though doubtfully, in the absence of fossils.   The association of these coarse-grained horn-blendic rocks with Arenig strata is seen not only here, but at Penarfynydd, numerous localities near Llanerchymedd and Am-lwch in Anglesey, and probably on Little Knott in Cumberland. Many of these masses have the general appearance of laccolites.

The intrusions near Clynog-fawr are doubtless the source of the numerous erratics which are scattered about in the neighbour-hood and to the south-west, as far at least as the base of Moel Penllechog.   Professor Sedgwick collected specimens from these boulders and from the Pen-y-rhiwiau mass itself, as well as a rock labelled "one mile N.E. of Clynog", which is probably from the dyke or mass near Tanrallt.   Professor Bonney[5] had slices cut

[1] *Min. Mag.*, vol. vi., p. 54; 1886.

[2] A. Geikie, *Trans. Royal Soc. Edin.*, vol. xxix., p. 506; 1880.   Teall, *Brit. Petr.*, p. 94, pl. vii.; 1888.

[3] Watts, *Report Brit. Assoc.* for 1887.

[4] Pen-y-rhiwau on the Survey map; Pen-y-rhiwan in Mr Tawney's paper.   On the maps of the Geological Survey the mass is carried a little too far to the east, and the Tanrallt intrusion is not marked.

[5] *G. M.*, 1880, pp. 457, 458.   *Q. J. G. S.*, vol. xli., p. 517; 1885.

from some of Sedgwick's specimens (Cl. 15 [132], Cl. 17 [102], Cl. 24 [133]), which he afterwards designated hornblende-picrite, and Mr Tawney[1] also examined the rock of Pen-y-rhiwiau. The general character of the mass hovers between hornblende-diabase and hornblende-picrite, olivine being either wanting or present in varying degree. If it be necessary to describe all the varieties under one name, that of olivine-hornblende-diabase conveys perhaps the best description of the average character.

Different specimens shew some differences in structure as well as constitution. The commonest rock is a coarse-grained and black one in which the eye recognises only large hornblende crystals with discoloured felspars and an occasional flake of golden-brown mica. The boulders, selected by a process of survival, are not typical of the whole mass, being mostly derived from the coarser portions of the outcrop. At Pen-y-rhiwiau the darker and more peridotic varieties occur chiefly near the west end of the mass, where the rock sometimes verges on the type of the Penarfynydd picrite. Many of the paler specimens from near the farm have little or no olivine. These bear the closest possible resemblance to the hornblende-diabases near Llanerchymedd in the centre of Anglesey[2], which I have described elsewhere. These latter too have furnished a plentiful supply of boulders scattered to the south-west, and shewing the same selection of the coarser and more basic variety.

Examined with a low objective, the sections are seen to consist mainly of amphibole in brown ophitic plates and green actinolitic crystals. The other constituents, however, are numerous.

Apatite occurs occasionally, but never abundantly, building slender needles or narrow hexagonal prisms, and being always the first product of consolidation [102]. A few grains of pale brown sphene are rarely seen [102]. Original iron-ores are sparingly present or entirely absent [102, 627]; but some slides shew magnetite in grains or in little octahedral crystals [586]. Secondary magnetite is an abundant product of the decomposition of hornblende and olivine. Copper pyrites [132] is occasionally present in imperfect crystals, and may be recognised in hand-specimens.

[1] G. M., 1882, p. 548.

[2] G. M., 1887, p. 546. See also Bonney, Q. J. G. S., vol. xxxvii., p. 137; 1881: vol. xxxix., p. 254; 1883: vol. xli., p. 515; 1885. Teall, Brit. Petr., pp. 81, 82, plates iv., vi.; 1888.

Olivine has probably been present in most of the rocks examined, though it is often completely destroyed. As already noted, its quantity is very variable [585, 587]. The alteration is of the usual kind, giving rise to serpentine with the mesh-structure and strings of magnetite granules or dust [586, 587]. Some calcite and dolomite are sometimes associated with the serpentine [102, 586].

Felspar occurs in variable quantity, being never a dominant constituent, and sometimes scarce or absent. It is too much decomposed into calcite, *etc.*, to afford any certain information as to its nature, but is presumably a variety rich in lime.

Unequivocal augite is mostly uncommon: possibly original augite has been amphibolised, and indeed the manner in which the grains are sometimes imbedded in brown hornblende, with the usual crystallographic relation, distinctly suggests this [586]. In one boulder, however—the only specimen containing much augite —it occurs in good crystals moulded by a similar brown hornblende. It is of a very pale yellowish brown tint or nearly colourless, with good cleavage and the octagonal cross-section due to the prism-faces and pinacoids [132]. It is often twinned on the orthopinacoid, and has an imperfect hour-glass structure.

A few colourless grains with good cleavage, imbedded in hornblende, may possibly be enstatite, and occasionally some part of the serpentine has a structure rather suggestive of derivation from a rhombic pyroxene; but this is far from certain. A similar mineral occurs in some of the Anglesey intrusions.

The ophitic plates of hornblende are, for the most part, deep brown, but in some places pass, either gradually or rather abruptly, into dull green or almost colourless [133, 627], sometimes with a zonary arrangement [132]. The absorption is expressible by the formula

$$\gamma \gtreqless \beta >> \alpha,$$

and the pleochroism by the following scheme:

Brown hornblende.	*Green hornblende.*	*Almost colourless.*
α, pale brown;	very pale rose-pink;	colourless;
β, deep clove brown;	pale olive green;	very pale grey-brown;
γ, deep clove brown;	rather pale grass-green;	very pale grey-brown.

The mineral usually shews perfect prismatic cleavage, and is sometimes twinned on the orthopinacoid [102], occasionally with repetition [627]. Rarely the prismatic cleavage is locally replaced by one parallel to the orthopinacoid [585]. In some of the boulders the hornblende shews crystal outlines—the prism and clinopinacoid [132, 133]. It sometimes [133] exhibits what appear to be ' solution-planes' parallel to the clinopinacoid, marked by discoloration and magnetite interpositions : a similar thing is seen in some of the Anglesey rocks [171, *etc.*]. Another common feature of these latter [535, 536, 539] is sometimes seen in the slides from Pen-y-rhiwiau. This is the 'secondary enlargement' of the original or pseudomorphic hornblende plates by a marginal growth of pale secondary amphibole-substance, often fibrous but in crystalline continuity with the compact hornblende. The same thing is well seen in the Little Knott rock [170], and has indeed been figured by Professor Bonney[1], though without any remark as to the secondary origin of the growth. It must be regarded as long posterior to the consolidation of the rock.

All the slides contain a considerable quantity of actinolite in rather pale grass-green blades, frequently with a fan-like arrangement. The crystals are pleochroic, the greatest absorption being for the γ-axis, which makes an angle of about 18° with the c-axis of crystallography. Some crystals shew a lamellar twinning, the twin-plane being the orthopinacoid [586]. The cleavage is apparently orthopinacoidal [585, 586], and there is an occasional cross-jointing [585]. The actinolite never includes olivine, but sometimes minute octahedra of magnetite. There can be little doubt that these green blades of actinolite are developed at the expense of the hornblende [585], and the passage already noted in this mineral from a prismatic to a pinacoidal cleavage appears to be an early stage of the transformation. Actinolite is much less abundant in the Llanerchymedd rocks.

Biotite occurs in irregular flakes, which sometimes give evidence of mechanical disturbance by shearing along 'gliding-planes' [133]. In ordinary light it is of a rather paler brown than usual, but there is the characteristically intense absorption for vibrations parallel to the cleavage-traces. This mica is sometimes an original constituent, being then newer than the augite, but older than the

---

[1] *Q. J. G. S.*, vol. xli., pl. xvi., fig. 2; 1885.

hornblende [132]. Most of the mineral, however, in the slides examined, seems to be secondary after the hornblende [586]. In some cases a partially disintegrated crystal of the latter mineral shews scales of biotite forming on its cleavage-planes; and, again, a mass of biotite is seen to enclose a kernel of decomposing hornblende [627].

Taking note of all the hornblendic rocks of North Wales, we may remark that the Clynog-fawr rocks are on the whole nearer to the Llanerchymedd hornblende-diabases than to those of the Penarfynydd and Rhiw district, and they are of interest as indicating an intermediate variety between the former and hornblende-picrite of the Penarfynydd type. The Tanrallt intrusion I have not examined so carefully as that of Pen-y-rhiwiau; but in hand-specimens and in a section [588] it exhibits very similar characters.

In various parts of Caernarvonshire occur a large number of dykes of sub-basic or sometimes basic composition, which may safely be separated from the Bala igneous rocks described in the foregoing pages, and assigned to a later age. They appear to be unconnected with visible plutonic masses or with volcanic outbursts properly so-called, and have a bearing usually between south-east and east-south-east, so that they cross the strike of the strata nearly at right angles. Many such dykes are noted on the Survey-map on the sea-coast and in the slate-quarries, and many others, in localities less freely exposed, may be found by searching. It may be observed that as we pass southward from the Menai Straits, the bearing of the dykes tends to be more easterly. Their width varies from a few inches to fifty or sixty feet.

The age of these dykes is a matter for inference only, but if we assume that they all belong to one period, they must be set down as post-Carboniferous, since many of them on the shores of the Menai Straits intersect the Mountain Limestone. The date may be more precisely fixed by comparison with the dykes of Anglesey [1]. Several of the Caernarvonshire dykes can be traced on the other

---

[1] *G. M.*, 1887, pp. 409—416; 1888, p. 267.

side of the Straits, and the dykes of this district thoroughly resemble others having the same direction in the Anglesey coal-field,
which latter are known with certainty to be post-Carboniferous but
pre-Permian [1]. They are sometimes injected in lines of fault connected with post-Carboniferous crust-movements. Such evidence
as we have, then, points to the probable supposition that all the
Caernarvonshire dykes here referred to belong to the interval
which separated the Carboniferous and New Red Sandstone periods
in this area.

The age of the dykes cutting the Llanberis slates seems at
first sight a more difficult question than that of the Menai Straits
dykes, with which they do not always agree in petrological character. But, in the first place, the strike of the dykes seems to
indicate their relation to the later folding ; secondly, Sir A. Ramsay
(pp. 102, 236) has pointed out that some of the dykes in Penrhyn
Quarry contain fragments of *cleaved* slate, proving the intrusion
to be later than the cleavage ; and, finally, the district shews no
manifestation of igneous activity between the close of the Bala age
and the close of the Carboniferous. The view that these dykes,
like the others, are post-Carboniferous seems, therefore, to be well-
founded.

The bearing of the dykes is, as already indicated, almost always
at right angles to the strike of the axes of post-Carboniferous
earth-movements in this district, and the igneous magma has
doubtless been injected into dip-joints. The persistent manner in
which the dykes maintain their course, both vertically and horizontally, must be explained here, as elsewhere, by the fact that the
pressure of the molten matter in a fissure must itself exert a powerful influence tending to rend the adjacent rocks apart in the direction perpendicular to the walls of the fissure, so that the dyke
constantly tends to propagate itself in its own plane, as was long
ago pointed out by Hopkins[2]. In the green schists between Porth-
dinlleyn and Bardsey the dykes run, on the whole, with the dip of
the strata, but it is quite in accordance with the difference between
the country-rocks that the regularity should be less marked here
than on the Menai Straits.

---

[1] Ramsay. pp. 205, 264. *Cf.* Sedgwick, *Q. J. G. S., Proc. G. S.*, vol. IV., p. 214;
1846.

[2] "Researches in Physical Geology," *Camb. Phil. Trans.*, vol. VI., p. 1.

The only published analysis of any of these rocks is one of a dyke in the Penrhyn Quarry by Dr Voelcker, quoted by Mr Maw[1], which gave the following figures:

SiO_2	47·47
TiO_2	2·51
Al_2O_3	5·80
Fe_2O_3	1·97
FeO	10·22
FeS_2	0·23
$CaSO_4$	0·08
$CaCO_3$	14·85
$MgCO_3$	14·59
K_2O	0·43
Na_2O	0·70
H_2O (combd.)	1·99
	100·84

The rock is evidently deeply altered, and indeed the carbonates were present as visible crystals. Eliminating the carbonic acid, *etc.*, the silica-percentage is raised to 56·24; but the other figures are still abnormal, and the analysis gives no clear idea of the original nature of the rock.

Mr Maw remarks how the blue slates are bleached in the neighbourhood of the dyke, and ascribes this circumstance to the reduction of most of the iron-peroxide by the heat consequent upon the intrusion. The slates at the junction with this and similar dykes are indurated, and their cleavage often impaired[2]—an additional proof of the post-Bala age of the dykes.

Petrologically these rocks are not all of one type, but gradations can be made out, and there seems to be no objection to treating them as a whole. Some of those found on the Menai Straits I have described in the paper already referred to[3].

The rock of the smallest dykes is invariably an augite-andesite of very fine texture, though no unindividualised basis can with certainty be detected. A hand-specimen, when fresh, is black

---

[1] *G. M.*, 1868, p. 125.

[2] *Cf. Brit. Assoc. Report* for 1885, p. 834.

[3] Numerous dykes in this district were recorded by Henslow, Trimmer, and Haughton in the papers quoted in our Introdution.

and compact, occasionally shewing clear felspars up to one-tenth of an inch in length with sometimes a fluxional disposition. Under the microscope these felspars shew albite and Carlsbad twinning, and appear to be labradorite or andesine. These felspars, with abundant crystals of magnetite, are imbedded in a ground-mass composed of felspar microlites, magnetite, and rounded augite granules. Olivine is not found. Dykes of this kind occur near Bangor, at Llanfair-is-gaer, on the shores of Llyn Padarn, and in other places. They are usually from four to eight inches in width. The same rock, however, often constitutes the marginal portion of larger dolerite dykes.

Many of the larger dykes are of dolerite of a moderately coarse grain, shewing even to the eye a well-marked ophitic structure. The microscope shews felspars of two generations, magnetite, and augite, with sometimes a little apatite. Ilmenite is not found, and olivine too is usually absent. The magnetite frequently builds rods or branching aggregates, and is commonly later than the first generation of felspars. The felspars, tested by their extinc-tion-angles, seem to range from labradorite to andesine, the later ones being more acid, as a rule, than the earlier. The earlier crystals give long rectangular sections with fine albite-lamellation, often combined with a further twinning on either the Carlsbad or the pericline law. The later felspars are less elongated and in general quite allotriomorphic. They have less close twin-lamella-tion than the others, and are further distinguished by a strongly marked zonary banding in polarised light, indicating by the test of extinction-angles that the outer zones are more acid in composition than the inner.

The augite shews in thin sections a pale brown tint, though not quite so pale as that of the common diabases of Caernarvonshire. In rare cases [736] there is a scarcely perceptible pleochroism, the colour varying from a rosy to a yellowish tinge. The augite forms ophitic plates moulding the earlier felspars and magnetite. It is sometimes slightly earlier, sometimes slightly later than the second generation of felspars: on the whole the two minerals are almost contemporaneous.

Dykes of this doleritic type are found at Craig-y-fael [737], Plas Rhiw [736], and other places in the Sarn district[1], as well as on

---

[1] *Cf. Q. J. G. S.*, vol. XLIV., pp. 460, 461; 1888.

the Menai Straits. Some near Bangor are visibly porphyritic in hand-specimens, shewing scattered squarish felspars, white, pink, or liver-coloured, half-an-inch in length. One of these dykes, at Glan Adda [527], contains a few scattered flakes of biotite, with fine needles of apatite, but as a rule mica is rare.

A few of the Menai Straits dykes are more distinctly basic than the foregoing, and contain rather abundant rounded grains of olivine. Such is the large dyke opposite Plas Newydd [529], which is a prolongation of the one on the Anglesey coast, and also a coarse-grained dyke exposed on the beach at Llanfair-is-gaer Church [559]. The latter contains a few flakes of biotite and some small patches of brown hornblende, included in the augite in the immediate vicinity of magnetite grains.

The dykes of the Penrhyn, Llanberis, and Nantlle quarries exhibit a considerable variety of characters, with gradations from the doleritic to the diabasic type. As an example of the former, we may take a specimen from north-west of Cwm-y-glo [560]. The slide shews small magnetite crystals, large and small lath-shaped felspars with twin-lamellation, a few shapeless later felspars with the usual zonary banding, and ophitic augite, rather abundant and later than any of the felspar.

A specimen from Llyn Padarn [561] is a diabase of rather 'granulitic' habit. It contains abundant frame-works of ilmenite, one generation of lath-shaped felspars, and plenty of the usual pale augite [1]. The last mineral, though occasionally penetrated by the felspars, is in rather rounded grains, often polysomatic. The abundance of ilmenite in the diabases and its exclusion from the typical dolerites are highly characteristic. It should be noted, however, that I adopt here, as elsewhere, the structural distinction between diabase and dolerite, the former having but one genera-tion of each constituent mineral, while in the latter there is a re-currence of one or more constituents, usually the felspars, giving the 'porphyritic' structure of Rosenbusch [2].

Some of the dykes contain a large quantity of epidote of secondary origin. A slice cut from a dyke nine feet wide in the Pen-y-bryn Quarry, Nantlle [562], affords a beautiful example of

---

[1] The "hornblende" of Ramsey (*Catalogue*, p. 46) is evidently augite.
[2] *Neu. Jahrb. f. Min. etc.*, 1882, vol. II., p. 1.

this mineral replacing felspar crystals by perfect pseudomorphs, but in the same rock are large nests replacing a considerable part of the whole mass.

The dykes which intersect the green schists between Porth-dinlleyn and Bardsey seem to belong for the most part to the doleritic type, though some, especially the smaller ones, appear to be andesitic. The ophitic structure is sometimes very strikingly developed. In a dyke between Nevin and Porth-dinlleyn [126] the augite plates not only mould and enclose the felspars, but even protrude little tongues along fissures produced by corrosion along cleavage-planes of that mineral. The same specimen shews a curious effect of crushing, possibly connected with the proximity of the boundary-fault. The rock is traversed by very numerous fine cracks filled with a pale-green feebly-polarising chloritoid substance[1]. These veins run parallel to one another across the slice, except in the interior of the larger felspars, where they are deviated so as to follow the cleavage-planes of the crystals.

These dykes in the green schists are assigned to a post-Carbo-niferous date necessarily with less confidence than the others enumerated above. It may be remarked, however, in support of this view that they have no apparent connexion with the intrusions in Lleyn which we have referred to the Bala age, and that un-doubted dykes of this age are decidedly rare in Caernarvonshire.

---

[1] This is the slide described by Mr A. S. Reid, *G. M.*, 1880, p. 457.

IT remains to be considered whether the facts brought forward in the preceding pages will serve as the material for any general conclusions with regard to the history of vulcanicity in North Wales during the Bala age. In this part of our subject we are necessarily on less firm ground than when merely recording petrological observations; but, as summarising and adding point to the detailed study of the rocks, a few speculations, advanced under due reserve, though prompted directly by the observed phenomena, may be considered not out of place. Something at least will have been gained, if we can shew reasons for regarding the igneous rocks of this area as not capricious and disconnected effects of forces acting under no special law, but, rather, closely related manifestations of organised processes, which may some day be reduced to intelligible principles. Many geological text-books, and even treatises on special districts, treat igneous rocks in a way which leaves the student with a general impression that they are of no particular age, but constitute meaningless interpolations in the geological record, to be dismissed in a brief appendix after the sedimentary formations have been duly discussed. The legends accompanying geological maps often serve to confirm this impression. The map illustrating a recent manual by a high authority indicates the volcanic rocks of Caernarvonshire and Antrim, of Iceland and the English lakes, by one common colour and legend.

The clue to the igneous phenomena of our district seems to lie in their relation to stresses operating in the crust of the earth during their production. Before proceeding to the evidences of this important relation, it will be necessary to consider the character of the movements which brought the Bala age to a close in the area in question.

That all the Bala and older rocks of North Wales were folded and cleaved before the deposition of the succeeding Silurian strata is a fact which seems to admit no dispute. An examination, for instance, of the tract bordering the lower part of the Conwy valley reveals at once the unconformity between the two sets of rocks, and further a certain discordance in strike between the feeble and variable cleavage of the newer strata and the highly-developed and constant cleavage of the older. The facts have been sufficiently enforced by Sir A. Ramsay. A comparative study of the directions of the cleavage-planes over the Caernarvonshire area enables us to realise in a simple manner the nature of the crust-movements which gave rise to the structure, and a glance at the chief undulations of the strata as marked on the Survey maps fully confirms the results so obtained. The succession of events which have followed as effects of the gradually augmenting lateral thrust may be clearly made out: (i) folding, (ii) cleavage, (iii) foliation and metamorphism. This is the chronological order, as might be anticipated on remembering the different degrees of yielding indicated by these changes: (i) a rearrangement of the position of the rock-masses as a whole, (ii) a rearrangement of their constituent fragments, (iii) molecular and atomic rearrangements. An examination of each of these three phenomena leads to the same conclusions as to the direction and distribution of the forces to which they owe their common origin. A glance at the map is enough to shew that these forces acted in lines having an average N.W.—S.E. direction. This, however, is not sufficient to define their action. All recent observations and theories alike lead us to regard lateral crust-movements as of the nature of a comparatively superficial creep of the rocks affected over those underlying, and therefore as taking place not only parallel to a definite line, but in a definite direction in that line. In Shropshire the movement was roughly from east to west; in the Lake District from south to north; in the eastern half of Caernarvonshire, as will be clearly seen in the sequel, the movement had a general direction from south-east to north-west.

Taking first the folding of the strata, and having regard more especially to the eastern division of the county, we note that the strike of the axes of disturbance is roughly N.E.—S.W., but curves in such a manner as to bear more nearly N.—S. towards the south-western limit of the district, and E.—W. in the north-eastern corner. In brief, the strike, where most clearly marked, shews on the map as a rather flat curve with its concavity facing the south-east, a circumstance in itself sufficient to suggest that the thrust came from that direction and not from the north-west. The straightness of the strike throughout a great part of its extent at once connects itself with the Llyn Padarn ridge of crystalline and other hard materials, to which the folds of the strata are closely parallel. This ridge appears to have constituted a comparatively firm buttress, against which the whole of the eastern division of Caernarvonshire was pressed, and its influence is traceable not only in the folding, but still more in the direction and relative development of the cleavage and local micro-foliation of the slates.

The Llyn Padarn ridge ranges in an approximately N.E.— S.W. direction from St Anne's Chapel near Nant Ffrancon to Llanllyfni, and must be imagined as prolonged under-ground to some little distance in both directions. This ridge appears to have played a most important part in the geology of Caernarvon- shire, and it will be frequently referred to below. Its antiquity is proved by the occurrence of abundant pebbles of its characteristic quartz-porphyry in the succeeding conglomerate, and the idea of its intrusion into the overlying rocks must therefore be con- sidered as finally abandoned. Most geologists who have studied the district regard the quartz-porphyry and its associated rocks as older than the whole of the Cambrian formations. Their Archæan age has indeed been called in question by Mr Blake, who con- siders them to be interbedded igneous rocks forming part of the Cambrian itself. The fact that the conglomerate appears to rest on both flanks of the ridge [1], and also the succession of the over- lying rocks, as exposed for instance in Nant Ffrancon, seem to me to support the Archæan theory. All this, however, is not material to our present purpose. The Llyn Padarn ridge existed as a buttress of relatively unyielding rocks during Bala and post-Bala times, and we shall see that it has exercised a ruling influence not only on the post-Bala folding and cleavage, but also on the manifestations of volcanic activity during the Bala age itself.

The folding of the strata in the eastern division of Caernarvon-

---

[1] The absence of the conglomerate near Cwm-y-glo and Pont-rhythallt is naturally explained by faulting, as shewn on the Survey map.

shire increases in intensity on the whole as we pass from south-east to north-west. This is better seen in the field than on the maps, for abstraction must be made of the *general* south-easterly dip, and moreover the most marked folding is on a small scale. The maximum is reached in what we may name the Llanberis slate-zone, using the term in a topographical not a stratigraphical sense. This zone, lying immediately on the south-east side of the Llyn Padarn ridge, is where the thrust from the south-east encountered most resistance. Here the sections in the slate-quarries shew an intense plication of the strata, and here too are seen examples of the local thickening of diabase dykes by the compression of folds under extreme pressure[1].

In most disturbed districts the folding is a less accurate index of the distribution of stresses in the earth's crust than that obtained from a study of the slaty cleavage. Here, however, the two correspond in strike not only as regards their average direction, but also in their chief deviations from the general bearing, thus proving that the plication of the strata is strictly a result of the same forces that produced the cleavage-structure. Mere upheaval and depression have had no share in determining the strike of the beds in this district, nor have subsequent accidents occasioned any material change in this respect. Indeed the post-Silurian and post-Carboniferous crust-movements seem to have had the same general direction as those now under discussion, and the same is true of the movements which affected the Archæan rocks before the deposition of the oldest Cambrian strata.

Over the greater part of the district the folding has not been accompanied to any considerable extent by faulting, and the great faults in the northern part of the county are clearly post-Carboniferous. Near the boundary between our eastern and western divisions, however, there are some interesting faults which must be assigned to the period of the folding, and these will be referred to below.

The arrangement of the cleavage-planes may be easily realised by means of the rough sketch-map (fig. 5) drawn up from a large number of observations. The lines indicate the strike of the cleavage, while their closeness together is designed to give an

---

[1] *G. M.*, 1889, pp. 69, 70.

idea of the relative development of the structure in different
localities. The general bearing of the cleavage-strike is about
N.E. and S.W., and this direction becomes more marked in proportion as we approach the Llyn Padarn ridge, which itself has
the same strike. Again, as we pass from south-east to north-
west across the eastern division of Caernarvonshire, the cleavage-
structure becomes more highly developed, and culminates on
reaching the Llanberis zone, on which are situated the famous
quarries of Bethesda, Llanberis, and Nantlle. In other words,
those rocks have been most crushed which lie nearest to the ridge
on its south-east side. The rocks composing the Llyn Padarn
ridge itself have been modified by the same forces, and the more
so on the side facing the thrust. Witness, for instance, the small
syncline extending north-easterly into the quartz-porphyry from
the middle of the lake. In this neighbourhood, too, the less
stubborn of the rocks constituting the ridge have received a
cleavage-structure equal to that of the neighbouring Cambrian
slates; while in some places, as near Pen-y-groes and Cilgwyn,
even the hard quartz-porphyry on the south-eastern face of the
ridge has been crushed into a schistose rock resembling the *por-
phyroïdes* of the Meuse valley. That the ridge yielded to some
small extent as a whole is proved by the slates on the other side
being cleaved, though far less perfectly than those in the quarries.
The structure, however, dies out rapidly as we travel north-west,
and the Arenig shales of Caernarvon and Bangor exhibit no trace
of cleavage. They have been protected by the Llyn Padarn ridge
from the powerful thrust which crushed, in varying degrees, all
the strata of the south-eastern division.

Over the greater part of the county the cleavage of the slates
is of the type so lucidly explained by Dr Sorby, and is not
attended by mineralogical changes. In the Llanberis slate-zone,
however, the case is different. The glossy slates or phyllites of
this tract have never been subjected to close study at the hands
of petrologists; but the microscopical examination of a few specimens is enough to confirm their general resemblance to the
schistes de Fumay so minutely analysed by Professor Renard.
They appear to consist in the main of quartz, white mica, chlorite,
and sometimes hæmatite, all presumably authigenic, and these
rocks must accordingly be ranked as highly metamorphic pro-

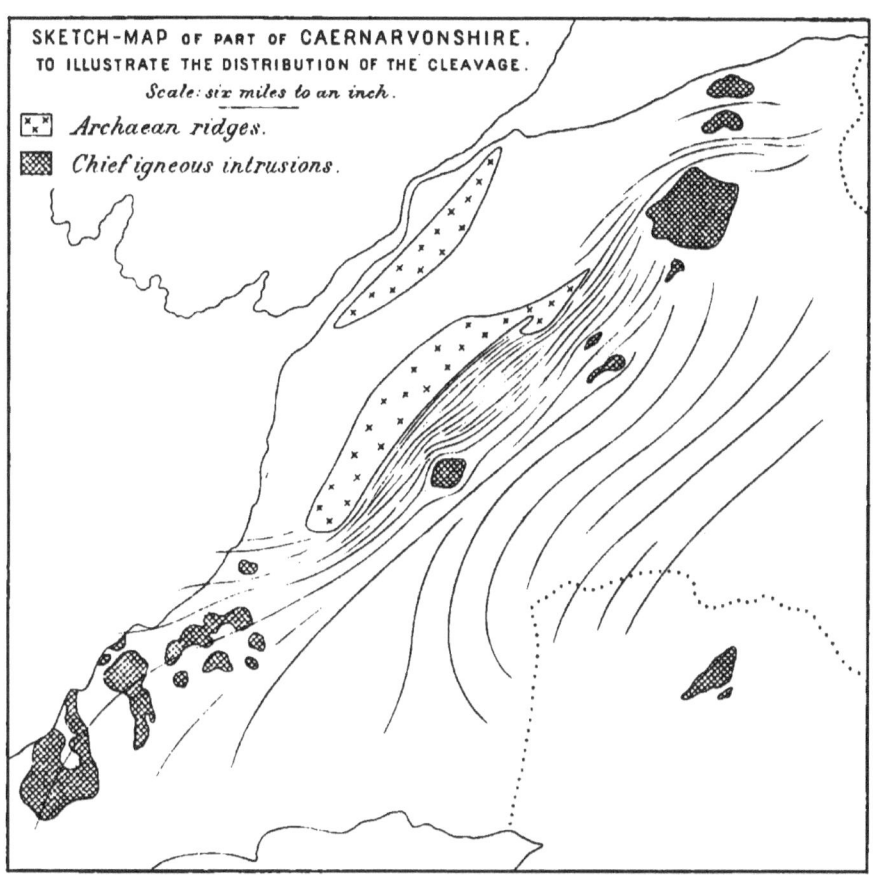

Fig.5.

ducts. It is significant enough that this micro-foliation is met with only in that tract which, as we have seen, was subjected to the most intense stress.

Certain secondary changes in the diabase sheets of eastern Caernarvonshire are possibly connected with the mechanical stress. Such are the production of epidote pseudomorphing felspar crystals, and of veins and borders of colourless fibrous amphibole in crystalline relation with the augite plates. These two features are seen in every slide cut from the diabases of the eastern division of the county, but never in those of the western, where, as we shall see, the lateral thrust was less powerful. Again, the ilmenite in the western diabases gives rise to the usual grey opaque leucoxene; but in the diabases of the eastern division the decomposition-product has a brown tint with a certain translucency and sometimes the double-refraction of sphene. This is a change which may very plausibly be referred to the dynamo-metamorphic effect of the more intense pressure in the latter case.

The great igneous sheets in eastern Caernarvonshire have exercised but little influence upon the direction or the quality of the cleavage in the adjacent slates. The solid plugs and bosses of igneous rocks at Mynydd Mawr[1] and Y Foel Frâs, on the other hand, have produced a very marked effect. The lines of cleavage-strike in each case may be seen to wind round the obstruction, as they would on a smaller scale round a hard imbedded nodule. The phenomena can be explained only on the supposition that the obstacle was there during the disturbance of the region, and are sufficient in themselves to establish the Bala age of the intrusions in question.

Passing now into the Lleyn district, we find the cleavage-structure much less strongly developed than in the eastern division of the county, and also shewing much less uniformity of strike. This is due to two reasons. In the first place, the Llyn Padarn ridge, which in the eastern district offered the requisite resistance to the thrust, and regulated all its effects, dies away when traced towards the south-west. But what is more important is that the direct thrust itself was in great measure confined to the eastern division of Caernarvonshire. To the west it had a diminished intensity and took in general a more westerly direction. Indeed

[1] *G. M.*, 1888, pp. 223, 224.

if we follow the lines of cleavage-strike in a south-westerly direction from the Llanberis slate-zone, we find them diverging like
stream-lines on escaping from a confined channel. On the one
hand they curve round westerly towards Clynog-fawr, owing to
the sinking of the Llyn Padarn ridge; and on the other they curve
southerly towards Tremadoc, owing to the diminished intensity of
the thrust and its more westerly direction in this part. By this
curving round of the 'stream-lines' the little tract to the north of
Tremadoc is made to occupy a peculiar position, being indeed, to
maintain the simile, a kind of eddy at the side of the stream.

To understand this fully it is necessary to realise the magnitude
of the horizontal displacement effected by the thrust in the eastern
division. The deformation of the green spots in the Llanberis
slates, first pointed out by Dr Sorby[1] and confirmed by many
other circumstances, proves that the strata occupying some parts
of that zone have been compressed in a nearly horizontal direction
into about one-fourth of their original horizontal breadth. The
compression in other parts of the district is of course less; but there
is no escape from the conclusion that the earth's crust in this
region has suffered a bodily horizontal displacement measurable in
some localities by miles.

The western division of the county has been much less affected,
and has, so to speak, lagged behind the eastern. In other words,
the eastern district has been relatively thrust forward past the
western. The bordering tract, to the north of Tremadoc, was thus
subjected to powerful shearing stresses acting horizontally along
vertical planes, and these stresses resulted in the formation of a
series of north and south faults. The faults, as laid down on Mr
Salter's map[2], seem to betoken exactly the kind of relative displacement here imagined. The modification of the rocks in this
tract is precisely in accord with the theory offered. Since the
intense stress induced was mainly a shearing stress, we should

[1] *Edin. New Phil. Jour.*, vol. LV., p. 137; 1853. *Cf.* Harker, *Report of Brit.
Assoc.* for 1885, pp. 813—852; 1886. Mr Fisher (*G. M.*, 1884, p. 402) has suggested
that these spots may be posterior to the cleavage. The regularity of their ellipsoidal
form is against this view; but what is more conclusive is that the spots are often
cut and displaced by small faults, which are quite obsolete as structural planes and
must themselves be older than the cleavage-structure. An instance is figured by
Mr Maw (*G. M.*, 1868, pl. vii.).

[2] *Catalogue of Coll. of Camb. and Sil. Foss.*, p. 9; 1873.

expect the argillaceous strata to yield to it without appreciable mineral change, while the more rigid diabases, resisting the deformation, would cause a conversion of the mechanical work of shearing into chemical energy, and so induce mineralogical transformations[1]. Accordingly we find that the slates of this neighbourhood offer no special peculiarity, but the diabases present much more decided evidence of dynamo-metamorphism than any others in Caernarvonshire. Secondary amphibole occurs not only in fibrous fringes and veins but in compact pellets with good hornblende-cleavage, and the titaniferous minerals have given rise to abundant clear grains of sphene, a feature not observed in any other Caernarvonshire diabases.

We are now in a position to realise the character of the folding and cleavage of the Caernarvonshire strata, and to some extent also the nature and distribution of the forces which brought about those phenomena. A much more detailed examination might easily be made, but would have no special application to our subject. The impression of the cleavage-structure, though doubtless in some degree a cumulative process, is as a whole clearly posterior to the plication of the strata. This is evident from the fact that the cleavage-dip is everywhere independent of the dip of the beds, and the fact is in agreement with what is seen in other areas of cleaved rocks. There can be no doubt, however, that the two phenomena are successive results of the same set of forces, and we are at liberty to regard these forces as growing gradually to a climax throughout a very long period of time. I hope to be able to shew some reason for considering the outburst of volcanic material during the Bala age as an earlier consequence of the same stresses in the earth's crust. It cannot be regarded as in itself improbable, that the growing forces should affect subterranean bodies of fluid, or potentially fluid, material, before they had gathered sufficient intensity to accomplish the folding of the solid rocks above.

Returning then to the igneous rocks, it must be noticed at the outset that their distribution presents striking relations with the position of the Llyn Padarn ridge, which seems to have played so important a part in the succeeding phenomena of folding and cleavage. Recalling our three-fold division of the county, it is of

[1] *G. M.*, 1889, pp. 18, 19.

sufficiently obvious significance that on one side of the ridge we find an ordered succession of lavas and intruded masses, but on the other side no igneous rocks of Bala age, with the possible exception of the Llanfaglen syenite; while in the Lleyn, where the hard axis is wanting, this simplicity of arrangement is entirely lost.

The next point which strikes the eye on glancing over a geological map of Caernarvonshire is that those intrusions of acid rocks, which have been indicated above as the centres of eruption, are situated very nearly on one line of strike drawn a little to the south-east of the Llyn Padarn ridge and its prolongation (see map, fig. 5). This is precisely the position they should occupy on the theory that the outbursts were due to a pressure from the south-east met by the resistance of the ridge. Such would be the line of weakness, through which at certain points the fluid lava would be forced, being lifted by the pressure into the sphere of action of volcanic agency, or perhaps actually squeezed out by this crust-pressure itself. Indeed the comparative scarcity of ashes seems to point to a steady, though intermittent, welling up of the fluid, with little intervention of explosive action; and the wide extent and great aggregate thickness of the main Snowdonian flows, apparently produced by simultaneous emission of lava from three or four vents, confirms this view. The individual flows, however, are never of extreme thickness, and the appearances lend no support to any idea of fissure-eruptions in this district. The several groups of flows into which we have divided the volcanic series shew a very evident distribution around certain centres, and these centres correspond to those intrusive masses which we have already been led by other considerations to regard as marking the probable vents. The sketch-map (fig. 6), designed to illustrate this distribution of the lavas, must of course be considered as rather a diagrammatic one. It is impossible to assign limits to the lavas with much accuracy, the continuity of the flows being so frequently lost, where they have been removed by denudation or are still thickly covered by newer rocks. It appears quite clear, nevertheless, that the Arenig lavas cannot extend far north of Moel-wyn and certainly not west of Tremadoc; that, of the Bala lavas, the three lowest groups must have come from one or more volcanoes in the neighbourhood of Y Foel Frâs; the uppermost group can be ascribed only to the Mynydd Mawr vent; while the main Snowdon group

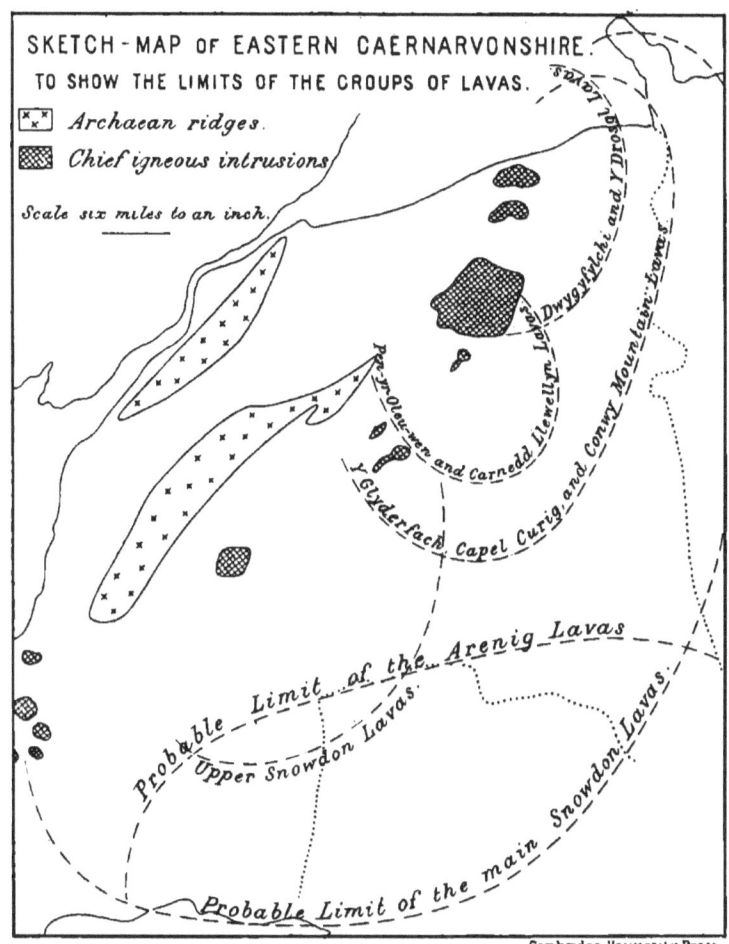

SKETCH-MAP OF EASTERN CAERNARVONSHIRE.
TO SHOW THE LIMITS OF THE GROUPS OF LAVAS.

Archaean ridges.

Chief igneous intrusions.

Scale six miles to an inch.

Cefni Lavas

Pen-y-Llithi and Y Drosgl Lavas

Llyn Dwygyfylchi and Conwy Mountain Lavas

Pen-Oleu-wen and Carnedd Llewellyn Lavas

Y Glyderfach Capel Curig Lavas

Probable Limit of the Arenig Lavas

Upper Snowdon Lavas

Probable Limit of the main Snowdon Lavas

Cambridge University Press.

Fig.6.

probably results from the commingling and overlapping of flows poured out from all the volcanoes which its range embraces.

There is one point of importance in connexion with the Bala volcanic series of Caernarvonshire which must not be lost sight of. It was long ago maintained by Sedgwick[1] that the whole of the lavas are of subaqueous origin, and the same view seems to be adopted by the author of the Survey memoir. Unfortunately the physics of lava-flows under water is a rather obscure subject. Scrope[2] considered that the molten material, owing to the pressure of the superincumbent water, will retain its included vapour, and therefore preserve its liquidity, for a long time, and will travel, under an instantaneously solidified crust, to greater distances than those reached by subaërial flows. A similar result was arrived at from various considerations by Daubeny[3] and others. The wide extent of some of the Caernarvonshire lava-flows and their general characters seem to be consistent with a submarine origin. It is by no means impossible that during some portion of the Middle Bala age the actual vents rose above sea-level, but the lavas themselves were probably for the most part solidified under the pressure of some depth of water. The alternative would require us to postulate a long succession of upheavals and depressions to account for the interbedded sedimentary strata. The volcanic material contained in the 'calcareous ashes' of Snowdon was probably ejected from a subaërial volcano, and fell into the sea to mingle with ordinary sediment and organic remains; but even this is by no means certain. If during the latter phases of their activity the volcanoes became elevated above sea-level, Caernarvonshire resembled in this respect the nearly coëval volcanic area of the English Lake District[4].

Turning now to the diabases, we see on any geological map that, as already pointed out, these rocks are for the most part intruded fairly accurately along the bedding-planes of the strata. It is true that most 'ideal sections' pourtray the sheets breaking across the strata and becoming irregular in form as soon as they disappear beneath the surface; but very little reflection will

---

[1] *Q. J. G. S.*, vol. III., p. 135; 1847.
[2] *Volcanoes*, pp. 244, etc. (2nd ed.), 1862.
[3] *A Description of active and extinct Volcanoes*, pp. 678, etc. (2nd ed.), 1848.
[4] J. Clifton Ward, *Mem. Geol. Surv.*, 101 S.E., p. 70; 1876.

make it clear, as already remarked in the case of laccolites, that when a number of intruded sheets crop out always with the same strike as the adjacent beds, it can only be because they are parallel to the planes of stratification. Excepting, then, such obviously transgressive masses as those about Carnedd Dafydd, Cwm-dyll, and Moel Siabod, with several outcrops in the Lleyn district, we may say that the normal mode of occurrence of the Caernarvonshire diabases is as 'sills,' *i.e.* intrusive sheets injected mainly along the bedding-planes.

We are naturally led to inquire the reason of this difference of behaviour between the acid and basic rocks of eastern Caernarvonshire, the one group being always extrusive, the other intrusive. Scrope[1] laid special stress on the "greater liquidity" of fused basic rocks "enabling them to penetrate the seams of stratified rocks among which they were forced." But it is by no means clear that the acid magma, with its content of aqueous vapour, was necessarily less fluid than the basic; nor that greater liquidity would cause a molten mass to force its way along bedding-planes rather than to the surface. Taking a different line, we may reasonably suppose, despite the rather extravagant speculations of Gilbert[2], that, in the molten as in the solid condition, the basic rocks have a higher specific gravity than the acid. On this the usual assumption it is not difficult to understand why the lighter rhyolites rose to the surface, and were poured out as lava-flows, while the heavier diabases were injected laterally into the strata without ever reaching the surface. The specific gravities of the two groups of rocks in the solid state are about 2·6 and 2·9 respectively.

Further, it appears highly probable that the diabasic magma came from a greater depth than the rhyolitic, and in fact underlay it in deep-seated reservoirs from which both appear to have been derived. The constant association of the two groups of rocks has already been commented upon, and it has been proved that the diabases, like the rhyolites, are of Bala age. Although the sheets of basic rock are individually newer than the rhyolitic *coulées* with which they are immediately associated, the two groups of rocks are on the whole coëval, and it is difficult to resist

[1] *Volcanoes*, 2nd ed., pp. 248, 249; 1862.
[2] *Geology of the Henry Mountains*, p. 70; 1880.

the conclusion that they are of common origin. The coincidence in their distribution, both in time and in space, seems too close to be explained by any hypothesis which makes the two magmas originally distinct and independent. We are therefore driven to the idea, which has been entertained by some geologists, of a process of separation under gravity taking place in the subterranean reservoirs which furnish the materials of both rocks. Such a theory—the reverse of Durocher's—pictures a molten rock-magma, originally of intermediate constitution, becoming separated under the action of gravity into upper acid layers of less density and lower basic layers of greater density. If such a theory be admissible, the different modes of occurrence of the rhyolites and the diabases in Caernarvonshire are completely accounted for.

In the English Lake District volcanic rocks of Bala age are largely developed, and the conditions of vulcanicity seem, despite some interesting differences, to have been very similar to those which obtained about the same time in Caernarvonshire. The Westmorland lavas, more intermixed with ashes and agglomerates than those in the Welsh area, divide generally into a lower andesitic and an upper rhyolitic series, and closely resemble the corresponding rocks described in the foregoing pages. The andesites may be supposed to represent approximately the original magma of intermediate composition, which was in part extravasated before the process of separation under gravity became effective[1]. In Caernarvonshire, on the other hand, it appears that the process of separation was in general well advanced prior to the outburst of any of the volcanoes. It is worthy of notice, however, that the andesites and andesitic agglomerates of Penmaen near Pwllheli are older than the rhyolites of that district, and the same is true of the other small exposures in Lleyn, if they are to be regarded as bedded lavas. Carn Boduan, if intrusive, is later than the acid rocks, and would seem to indicate a fresh accession of the original intermediate magma. To determine the true relations of the andesites of the Y Foel Frâs tract would require

---

[1] The general succession referred to of intermediate followed by acid rocks is best seen in Westmorland. It should be noticed that basic lavas are not wanting in the Lake District: see analyses of specimens from Eycott Hill, Cumberland; J. C. Ward, *Monthly Microscopical Journal*, vol. xviii., p. 246; 1877.

detailed mapping combined with the examination of a larger number of specimens than I have found opportunity to collect[1]. The agglomerates at Llyn-yr-afon can scarcely be intruded through the neighbouring granophyres, as the andesites of Bera-mawr might be supposed to be. From the manner in which the great boss of granophyre has encroached upon the lowest lavas, whose source must have been somewhere within the tract in question, we are justified in supposing that an earlier volcano in this vicinity has been destroyed by a later and more extensive invasion of the igneous magma. Other facts support this idea. The quartz-dolerites and andesites of Penmaenmawr and Tai-rhedyn, which may well be relics of one larger mass, also cut into the lowest lavas, and presumably represent a laccolitic intrusion of inter-mediate rock attendant upon the later outbreak of activity in the Y Foel Frâs centre. From the position of the laccolite in the arch of an anticlinal fold, it even seems likely that when the intrusion was effected, a certain degree of folding was already in progress in this extreme northern part of the county.

If we imagine a large reservoir of molten rock, already well separated into layers of different densities, to find exit above in obedience to increasing pressure within, it seems evident that, by a process of decanting, the upper and lighter layers will first be ejected, then the lower and heavier layers in order; so that the lavas poured out will be each less acid than the preceding. On the other hand, while the differentiation of the original magma is still incomplete, the upper layers of our imaginary reservoir may be becoming progressively more acid, and must do so when the extravasation of material from the vent above is checked by some relief of the internal pressure. In this case the newer lavas will be more acid than the older. This latter set of conditions must be the more usual, and would probably be verified to some extent by a systematic series of analyses of lavas in the Caernarvonshire area. The lower groups are evidently less acid in constitution than the main Snowdon lavas, and, from such analyses as we can command, this group itself seems to grow more acid from the base upwards[2]; while the small group of flows which we have named

---

[1] Three attempts to traverse these mountains have been frustrated by the snow.

[2] In the Cader Idris district Messrs Grenville Cole and Jennings find that the

the upper Snowdon lavas, by a falling off in acidity, indicate the exhaustion of the lighter portions of the subterranean magma.

The cessation of superficial volcanic action in any particular tract of the area must then be held to signify, not a relaxation of the crust-pressure to which all the phenomena are here referred, but a change in the mode of its manifestation. Instead of an ejection of the lighter magma there succeeded an injection of the heavier. The field-geology of the area sufficiently proves, as I have already insisted, that the diabasic intrusions followed in general the rhyolitic extrusions, and in the same tracts; that the material came from greater depths; and that when some of the diabases were being injected, certain parts of the area had already begun to experience a folding of their stratified rocks.

The laccolitic intrusions of the most basic parts of the separated rock-magma afford evidence completely in accord with the ideas here put forward. These rocks, viz. the hornblende-diabases and picrites, are met with only in parts of the area where highly acid rhyolites attest a very perfect differentiation of the original magma, and they occupy a very low horizon, being found invariably in strata which have been referred to the Arenig series, and far below the horizon of any extruded lavas in Caernarvonshire. The high specific gravity of these rocks seems to have prevented their breaking upward through the strata even to the extent that the diabases have been able to do so.

We have already had occasion to remark the want of igneous rocks of Bala age on the north-west side of the Llyn Padarn ridge. As regards the lava-flows, this may of course be explained by denudation, but the complete absence of acid intrusions, diabase sheets, and injections of hornblendic rocks in this division of the county admits of no explanation but one similar to that advanced, taking account of the ridge itself as a barrier effectively dividing this district from eastern Caernarvonshire. The Llanfaglen rocks may conceivably be due to a stray intrusion of intermediate rock-magma connected with the igneous rocks of the other side of the ridge, but there is nothing to definitely fix their age. Being post-Arenig and having obviously no affinity with the post-

tuffs and ashes become more acid from the base upwards. *Q. J. G. S.*, vol. xlv., pp. 436, 437; 1889.

Carboniferous intrusions of the district, they may be placed provisionally among the Bala rocks.

It is not necessary to suppose that the barrier was absolutely immoveable. We have seen that at a later time it yielded somewhat to the augmented thrust from the south-east. Possibly during the volcanic age itself the ridge—then covered with a great thickness of Cambrian (and Ordovician) strata—may sometimes even have risen above sea-level, the volcanoes fringing its unstable and oscillating shore-line. This would best explain some of the pyroclastic accumulations, and such an elevation would not leave any more decided traces of unconformity than we frequently see in the district in the local coalescence of distinct lava-flows separated elsewhere by sediments. A general elevation in the neighbourhood of the Llyn Padarn ridge during the period of the main Snowdon lavas, if sufficient to produce an appreciable slope of the sea-bottom, would help to explain the great distances to which some of the flows extended. In this case, the country to the north-west of the ridge may never have been covered by so great a development of lavas as is seen in the heart of the county.

In order to connect these various speculations arising from the study of the volcanic series of Caernarvonshire, it may not be out of place to give a hypothetical sketch of the course of events in this area during the Bala period.

With the close of the Arenig age, the volcanoes which had flooded the Merioneth area with their lavas, and sent a few flows over the Caernarvonshire border, became extinct, and they have probably had no revival. During the time immediately succeeding, there was a cessation of igneous activity, and when the volcanic fires again broke out it was in a new area, that of Caernarvonshire. The transference of the seat of action in this direction seems to agree with the nature of the lateral thrusts which have been discussed above. As already pointed out, there is good reason to suppose that in central and eastern Caernarvonshire the thrust was directed in general from south-east to north-west, but this was doubtless part of a divergent system of forces which took a more northerly direction in the north of the county and a more westerly direction in the district farther west. The rectilinear character of the folds subsequently formed in the central tract and the straight strike-line of the accompanying cleavage we have referred to the influence of the Llyn Padarn ridge; and it seems probable that this influence began to make itself felt at an early time in the Middle Bala age, being perhaps assisted by a certain amount of elevation of the Llyn Padarn axis as one of the first effects of the thrust.

The first outbursts of lava in the eastern half of Caernarvonshire were localised in the northern part. A volcano situated immediately north of Y Foel Frâs was probably the source of the lowest flows—those of Penmaenbach, Dwygyfylchi and the ridge of Y Drosgl. The magma erupted was of much less acid constitution than the typical rhyolites of Snowdon, *etc.*; but the next lavas, those of Pen-yr-Oleu-wen and Carnedd Llewellyn, shew that the differentiation of the subterranean magma was steadily progressing. As in the former case, the vent of these lavas must have been situated within the area now occupied by the large composite igneous mass of Y Foel Frâs, Bera Mawr, *etc.*, and the flows were of limited extent. After a pause, a more extensive outpouring of lavas followed in the same northern district, the relics of which may be traced from Conwy Mountain to Capel Curig and Y Glyder Fach. The distribution of these lavas unmistakably points to the Y Foel Frâs tract as the site of the volcano from which they were erupted; while their extent over a distance of fourteen miles and the number and thickness of the flows seem to indicate that the volcano was one of considerable size. It is to this time, therefore, that we may most probably refer the breaking out of the large Y Foel Frâs volcano, which destroyed the remains of the older vent, or perhaps the two older vents, from which the earlier lavas had issued. Meanwhile, under the augmented thrust, an incipient anticline had been formed running eastward from Llanfairfechan, and into the arch of this had been injected as a rude laccolite a portion of the new intermediate magma, prior to any effective separation under gravity. In this way had been formed the intrusion of Penmaenmawr and Tai-rhedyn, which, like the new volcano itself, is seen to encroach on the older lavas of Y Drosgl, *etc.*

During this time it is probable that large reservoirs of molten rock-magma under the districts farther south-west had been undergoing a process of separation into lighter and heavier strata;

so that the earliest products of the vents by which they found relief were very distinctly acid. It is evident indeed that considerable quantities of liquid rhyolitic material must have accumulated by the beginning of the next phase of vulcanicity, to which are due the main Snowdonian lavas. In the central district a portion of this acid material was injected among the sedimentary strata, forming the laccolites of Moel Perfedd and Bwlch-cywion, and here too a volcano was formed at what is now the head of Nant Ffrancon. From the slight interval which in this neighbourhood divides the flows of the main Snowdon group from those preceding, it would appear that this vent was the first to become operative during the new phase of activity; but there can be little doubt that the Y Foel Frâs volcano subsequently resumed action, while a large new one broke out, the plug of which now forms Mynydd Mawr. The number of flows gives evidence of a prolonged period of activity, while the thickness of some of them and the fact that they may be traced over a distance of more than thirty miles from north-east to south-west prove that the eruptions attained very large proportions, and were in part synchronous from the different vents.

There can be no doubt that eruptions in the Lleyn peninsula were more or less contemporaneous with those that gave birth to the main Snowdon lavas in the east, but it would be difficult to fix the exact time at which volcanic outbursts first took place in this south-western division of the county. We know that the most important rhyolites were there preceded by some small andesitic flows, perhaps fairly representing the undivided intermediate magma. The situations of the vents must be sought among such intrusive masses as those near Clynog-fawr, Yr Eifl, and Carn Fadryn.

When the volcanoes of the eastern division became extinct, the last to die was that of Mynydd Mawr, to which must be assigned the uppermost lavas. These were divided from the main Snowdon group by an interval in which the only volcanic products were of a fragmental character, and with these feeble flows the last phase of superficial vulcanicity came to a close. Most likely eruptions had ceased in Lleyn also, being succeeded there by sporadic intrusions of acid magma, which gave rise to rhyolite, quartz-porphyry, granophyre, or even granite, according to the conditions of consolidation. Possibly Carn Boduan, apparently an old vent plugged with andesitic lava, represents a late volcano arrested at an early stage of its progress; but the denudation which removed the higher Bala beds from this part of the county, has left the history here incomplete.

The outpouring of acid lavas being completed, the remaining heavier portion of the original magma was injected in the form of diabase sills and occasionally bosses. This may have been going on to some extent during the period of the volcanoes, but on the whole the diabases are obviously newer than the rhyolites. The strata began to yield more and more to the still increasing thrust from the south-east, and in the east of the county the injection of the diabases proceeded concurrently with the plication, the molten mass finding its most easy channel along the axes of the folds. The basic hornblendic rocks, intruded on low horizons, probably belong to a rather late phase of the igneous action.

Meanwhile the Llyn Padarn ridge had risen into a more formidable barrier, and other features, such as the Cwm Tryfan and Capel Curig anticlines and the Snowdon, Cefn-y-capel, and Dolwyddelen synclines, had declared themselves. These were intensified by the growing thrust; the cleavage of the more yielding beds accompanied and followed the severe plication of the strata; and mineral rearrangements were produced in the tracts where the most intense stresses were localised. Finally, the whole county having become elevated above sea-level, the rocks were subjected to ordinary denuding agencies, and the lateral thrusts, which had reached a climax during the cleavage and foliation of the slates, entirely died away. It is impossible to determine how much, if any, of the Upper Bala formation was ever deposited in Caernarvonshire; but the appearances seem to indicate that the break between the volcanic series and the succeeding Silurians of the eastern border is one of considerable magnitude, and it is possible that this county had already risen bodily above the water, while marine deposits were still forming in the country farther south.

In conclusion, a few words on stratigraphical grouping may not be inappropriate. The question is one scarcely germane to our subject, although it is possibly touched by the title originally proposed for this essay.

H. E.   9

Any classification to be widely applied must be founded on the succession of palæontological forms. The existence of three faunas in the Lower Palæozoic rocks was clearly admitted in the discussion at the London Congress, and the general adoption of three distinct names for the three systems of rocks is probably only a question of time.

The physical break at the summit of the Bala in this the Cambrian district of Sedgwick has been emphasised in the foregoing pages, and has been ascribed to the same general causes as the vulcanicity of the Bala age. It will be quite in accord with this theory if further investigation should reveal a break, though one of less magnitude, above the Arenig rocks on the southern borders of Caernarvonshire. The Ordovician period, including the Arenig and Bala (with Llandeilo), is distinguished physically in North Wales from the immediately preceding and succeeding periods by the prevalence of volcanic manifestations[1]. Since the Bala volcanoes became extinct, North Wales has known no important outbreak of igneous activity. In the interval between the Carboniferous and Permian periods, when molten rock rose in numerous fissures within the Caernarvonshire area, we have no evidence of any superficial demonstration of the subterranean forces; while the great Tertiary outbursts which convulsed so large a portion of the British Isles did not extend into the principality. The history of vulcanicity in North Wales is the history of the pre-Cambrian and the Ordovician periods; and when this area comes to be studied in detail by abler geologists, we may hope to learn from it at least as much of the mechanism of igneous action and the internal economy of volcanoes as any other district, Palæozoic or Neozoic, is able to teach us.

---

[1] Messrs Grenville Cole and Jennings, however, place the earliest eruptions of the Cader Idris district as far back as the Tremadoc age or even earlier; while Mr Blake regards the Llyn Padarn ridge as evidence of volcanic eruptions during the Harlech and Llanberis age.

CAMBRIDGE: PRINTED BY C. J. CLAY, M.A. & SONS, AT THE UNIVERSITY PRESS.